CHEMISTRY RESEARCH AND APPLICATIONS

PROPERTIES AND USES OF BUTANOL

CHEMISTRY RESEARCH AND APPLICATIONS

Additional books and e-books in this series can be found on Nova's website under the Series tab.

CHEMISTRY RESEARCH AND APPLICATIONS

PROPERTIES AND USES OF BUTANOL

ARNAUD M. ARTOIS
EDITOR

Copyright © 2020 by Nova Science Publishers, Inc.

All rights reserved. No part of this book may be reproduced, stored in a retrieval system or transmitted in any form or by any means: electronic, electrostatic, magnetic, tape, mechanical photocopying, recording or otherwise without the written permission of the Publisher.

We have partnered with Copyright Clearance Center to make it easy for you to obtain permissions to reuse content from this publication. Simply navigate to this publication's page on Nova's website and locate the "Get Permission" button below the title description. This button is linked directly to the title's permission page on copyright.com. Alternatively, you can visit copyright.com and search by title, ISBN, or ISSN.

For further questions about using the service on copyright.com, please contact:
Copyright Clearance Center
Phone: +1-(978) 750-8400   Fax: +1-(978) 750-4470   E-mail: info@copyright.com.

NOTICE TO THE READER

The Publisher has taken reasonable care in the preparation of this book, but makes no expressed or implied warranty of any kind and assumes no responsibility for any errors or omissions. No liability is assumed for incidental or consequential damages in connection with or arising out of information contained in this book. The Publisher shall not be liable for any special, consequential, or exemplary damages resulting, in whole or in part, from the readers' use of, or reliance upon, this material. Any parts of this book based on government reports are so indicated and copyright is claimed for those parts to the extent applicable to compilations of such works.

Independent verification should be sought for any data, advice or recommendations contained in this book. In addition, no responsibility is assumed by the Publisher for any injury and/or damage to persons or property arising from any methods, products, instructions, ideas or otherwise contained in this publication.

This publication is designed to provide accurate and authoritative information with regard to the subject matter covered herein. It is sold with the clear understanding that the Publisher is not engaged in rendering legal or any other professional services. If legal or any other expert assistance is required, the services of a competent person should be sought. FROM A DECLARATION OF PARTICIPANTS JOINTLY ADOPTED BY A COMMITTEE OF THE AMERICAN BAR ASSOCIATION AND A COMMITTEE OF PUBLISHERS.

Additional color graphics may be available in the e-book version of this book.

Library of Congress Cataloging-in-Publication Data

Names: Artois, Arnaud M., editor.
Title: Properties and uses of butanol / Arnaud M. Artois, editor.
Description: New York : Nova Science Publishers, [2020] | Series: Chemistry
   research and applications | Includes bibliographical references and
   index. |
Identifiers: LCCN 2020035221 (print) | LCCN 2020035222 (ebook) | ISBN
   9781536184488 (paperback) | ISBN 9781536185447 (adobe pdf)
Subjects: LCSH: Biomass energy. | Butanol. | Fermentation.
Classification: LCC TP339 .P77 2020 (print) | LCC TP339 (ebook) | DDC
   662/.88--dc23
LC record available at https://lccn.loc.gov/2020035221
LC ebook record available at https://lccn.loc.gov/2020035222

Published by Nova Science Publishers, Inc. † New York

Contents

Preface		**vii**
Chapter 1	Properties and Uses of Butanol *Rupesh K. Gautam, Anjali Saharan,* *Kashish Wilson and Komal Preet Kaur*	**1**
Chapter 2	Production of Butanol from Biomass: Advances and Challenges *Fernando Israel Gómez-Castro,* *Eduardo Sánchez-Ramírez,* *Ricardo Morales-Rodriguez,* *Juan José Quiroz-Ramírez* *and Juan Gabriel Segovia-Hernández*	**27**
Chapter 3	Butanol Oxidation Reaction: From Pt Single Crystal to Direct Butanol Fuel Cell *Guilhermina F. Teixeira, Tarso L. Bastos,* *Enrique Herrero, Juan M. Feliu* *and Flavio Colmati*	**101**
Index		**151**

PREFACE

Properties and Uses of Butanol reviews the different types of butanol along with its characteristics, methods of production and future trends observed in its applications as an alternative energy resource.

The main aspects involved in the production of biobutanol are described, including raw materials, the transformation of biomass and the separation of the acetone-butanol-ethanol mixture. The most important areas of opportunity are determined, focusing on the enhancements required by the production process to increase reaction yields in the hydrolysis and fermentation steps.

The closing study discusses the oxidation of butanol on Pt single crystal, the possible mechanisms of the butanol oxidation reaction, and the working principles of fuel cells.

Chapter 1 - Due to increase in energy crisis leading to a great demand for the alternative sources of energy as of fossil fuels are going to be extinct day by day. The evolution of new technology which was used for the production of fuels and chemicals from butanol will makes it importance in the new upcoming era to maintain and get rid of current global energy crisis. n-butanol is a versatile and sustainable platform chemical that can be produced from a variety of waste biomass sources such as sugar corns, food wastes and fruit residues. Butanol was also found to have similar characteristics to gasoline and can be used as alternative oxygenate to ethanol in blended fuels. Butanol can be substituted to be as better fuel than ethanol due to characteristic features that is lower vapor pressure and higher energy density. In the last few decades it was evidenced that major improvements and changes were reported in the production of fuels and chemicals by using various methods such as by using fermentation, bioreactors, enzymes and microbes such as clostridia, genetically engineered *Escherichia coli (E.coli) and Saccharomyces cerevisiae* in reactions such as ABE method of fermentation (acetone butanol ethanol) and corn steep liquor. Apart from it butanol also depicts different isomeric forms with a variety of applications in different field such as an organic solvent, plasticizer, cosmetics and in gasoline. In the present chapter the authors cover the different types of butanol along with its characteristics, methods of production and future trends observed in the same as alternative energy resource.

Chapter 2 - Butanol is an alcohol that can be used as fuel in internal combustion engines as a replacement for gasoline. Even though ethanol is the most known alcohol for such use, it is commonly employed as an additive instead of a fuel, since it cannot be used 100% in the existing engines. On the other hand, butanol has similar properties to those of gasoline, turning it in a better alternative to run vehicle engines. Like ethanol, butanol can be produced from sugar/starchy biomass and lignocellulosic biomass. Butanol produced from such raw materials is known as biobutanol. The traditional production process for that product is known as ABE fermentation since acetone, butanol, and ethanol are obtained. When lignocellulosic materials are used as raw material, pre-treatments are required before the fermentation step. Although ABE fermentation is a relatively known process, there are still several challenges to turn biobutanol into an economically competitive product. In this chapter, the main aspects involved in the production of biobutanol are described, including raw materials, the transformation of biomass and the separation of the ABE mixture. The most important areas of opportunity are described, focusing on the enhancements required by the production process to increase reaction yields in the hydrolysis and fermentation steps; and to reduce costs and reduce environmental impact in the purification trains.

Chapter 3 - Most part of the global energy is coming from fossil resources. Besides not being a renewable energy source, the products of their combustion processes are harmful to the environment. Fuel cells are a great renewable alternative to be used in energy production. The electricity generated by fuel cells is the result of the direct energy conversion promoted by the electrochemical transformation of compounds that store large amounts of hydrogen. Alcohols are often used as hydrogen sources and the chemical reactions that lead to the power generation can be catalyzed by precious metals. Platinum-based catalysts are efficient materials that can be used in the fuel cell operation. To synthetize the Pt catalysts, the most usual procedure consists of the reduction of soluble Pt species to produce nanoparticles with different morphology. The catalyst shape control is an important factor to consider since it results in atomic arrangements on the surface of the particle that leads to different exposed Pt active sites. These characteristics heavily depend on the synthetic method of the nanoparticles. After an adequate surface characterization, it is possible to establish a reliable mechanism for the oxidation reaction of hydrogen-rich compounds to generate energy and to find catalysts with higher activity for this reaction. Based on these contexts, in this chapter, the authors will discuss the oxidation of the butanol on Pt single crystal, the possible mechanism of the butanol oxidation reaction, and the working principles of fuel cells as well as the advantages of employing the Pt-based nanoparticles as catalysts in the energy generation process.

In: Properties and Uses of Butanol
Editor: Arnaud M. Artois
ISBN: 978-1-53618-448-8
© 2020 Nova Science Publishers, Inc.

Chapter 1

PROPERTIES AND USES OF BUTANOL

Rupesh K. Gautam[1,], Anjali Saharan[1], Kashish Wilson[2] and Komal Preet Kaur[1]*

[1] MM School of Pharmacy, Maharishi Markandeshwar University, Sadopur-Ambala- Haryana, India-134007
[2] MM (Deemed to be University), Mullana-Ambala-India

ABSTRACT

Due to increase in energy crisis leading to a great demand for the alternative sources of energy as of fossil fuels are going to be extinct day by day. The evolution of new technology which was used for the production of fuels and chemicals from butanol will makes it importance in the new upcoming era to maintain and get rid of current global energy crisis. n-butanol is a versatile and sustainable platform chemical that can be produced from a variety of waste biomass sources such as sugar corns, food wastes and fruit residues. Butanol was also found to have similar characteristics to gasoline and can be used as alternative oxygenate to ethanol in blended fuels. Butanol can be substituted to be as better fuel than ethanol due to characteristic features that is lower vapor pressure and

---

*Corresponding Author's Email: drrupeshgautam@gmail.com.

higher energy density. In the last few decades it was evidenced that major improvements and changes were reported in the production of fuels and chemicals by using various methods such as by using fermentation, bioreactors, enzymes and microbes such as clostridia, genetically engineered *Escherichia coli (E.coli) and Saccharomyces cerevisiae* in reactions such as ABE method of fermentation (acetone butanol ethanol) and corn steep liquor. Apart from it butanol also depicts different isomeric forms with a variety of applications in different field such as an organic solvent, plasticizer, cosmetics and in gasoline. In the present chapter we cover the different types of butanol along with its characteristics, methods of production and future trends observed in the same as alternative energy resource.

Keywords: butanol, biofuel, ABE fermentation, corn steep liquor (CSL), production

ABBREVIATIONS

E.coli	Escherichia coli
Butyryl-co-A	Butyryl-coenzyme A
Butyryl-co-B	Butyryl-coenzyme B
CSL	Corn steep liquor
ABE method	Acetone- butanol- ethanol fermentation
Aceto-acetyl-Co-A	Acetoacetyl Coenzyme A
ATP	Adenosine triphosphate
NADPH	Nicotinamide adenine dinucleotide phosphate
Tert-butanol	Tertiary butanol
C.acetobutylicum	Clostridium acetobutylicum
UV	Ultra Violet
DNA	Deoxyribonucleic Acid
C. Beijerinckii	Clostridium beijerinckii
HPLC	High performance liquid chromatography
OH	Alcohol

1. Introduction

An increasing demand of the fossil fuels creates a tremendous pressure on the existing resources. Due to high consumption and less production rate leads to emergence of research in new field area such as biofuels. Now a day, biofuels becomes a vital component of our day to day life in modern era. Bioethanol components are new emerging sources of fuels which are in great demand due to presence of ethyl esters fatty acids present on them. Due to presence of these groups they are widely used in the production of gasoline and diesel fuels. Among various forms of biofuel the butanol was found to be more productive with variant properties and beneficial effects [1].

Butanol is a four membered carbon chain with alcohol moiety present at terminal positions. Butanol was found to be a clear neutral liquid with a strong characteristic odour. It is mainly miscible with majority of the solvents such as alcohols, ether, aldehydes and ketones, aliphatic and aromatic hydrocarbons which is sparingly soluble in water with a high and is highly refractive compound. Butanol is a four chain carbon atom with alcohol which exists in four different isomeric forms such Primary butanol or n- butanol, Secondary butanol, Tertiary butanol and isobutanol. It was found that Primary butanol and Isobutanol have the characteristic property which makes them effective for use as biofuels. The mode of production of these four isomers differs as only n-butanol, secbutanol and isobutanol are produced by microbes whereas tert-butanol is mainly found and produced in refinery. Butanol is the only fuel which is produced by microorganism which can be utilized as fuel or fuel component. Butanol was also known to be easily miscible with higher hyrdocarbons including gasoline due to its long chain property [2].

Butanol is produced chemically mainly by two methods such as by oxo process which started from propylene along with H and Co are used as catalyst over rhodium and second process is aldol process starting from the acetaldehyde. Butanol is a well-known chemical compound with super power fuel characteristics which contain the oxygen content of around 22%, which was used for complete combustion of fuel [3]. Butanol also found to be effective and versatile as it contributes in fresh and clean air by reducing emission of smog creating compounds, harmful emissions such as carbon monoxide and unburned hydrocarbons in the tail pipe exhaust. The petrobutanol was formed by chemical synthesis of fossil fuels. On the other hand biobutanol which is formed by microbial fermentation of biomass [4].

Butanol is less corrosive in nature due to the lower value of hygroscopic profile. Due to this reason it can be directly used within engine with no major modifications because it got easily blend with gasoline at different ratios and can be easily transported. Butanol also plays an important role as a precursor in plastic and polymers production. It is also widely used as solvent in the production of antibiotics, hormones, vitamins in the textile industry and as a perfume base in the cosmetics. As a conclusion of above mentioned characteristics and features, butanol was found to be the most promising among the conventional fuel alternatives because of its uncountable advantages over the other known biofuels [5].

2. Characteristic Properties of Butanol

Table 1. Comparison of different characteristics of Butanol over Ethanol

Characteristic	Ethanol	Butanol
Formula	C_2H_5OH	C_4H_9OH
Boiling point (°C)	78	118
Energy density (MJ Kg^{-1})	26.9	33.1
Air fuel ratio	9.0	11.2
Research octane number	129	96
Motor octane number	102	78
Heat of vaporization (MJ Kg^{-1})	0.92	0.43
Vapour pressure	112-129	96
Calorific value	26.8	32.5

Butanol as biofuel possesses several advantages over the other fuels due to the characteristic properties shown by its nature such as density, heat of vaporization, pressure etc. As a result of these characteristic makes its advantages over the other fuels. The characteristic properties of the butanol is compared with ethanol in Table 1.

2.1. Heating Value

Alcohol was found to possess low heating value and its rises with increased carbon content of it. n – Butanol possess more energy along with it was found to have high density and volume because of the four carbon moieties [6]. Due to these properties butanol was found to be less consumed and highly productive in terms of mileage when compared to ethanol.

2.2. Ignition Consumption

In comparison to ethanol, vaporization heat is also on lower level in n-butanol.

2.3. Inter-Solubility

n- Butanol was found to possess excellent gasoline solubility inert measures. Non-polar hydrocarbons are less polar as well as less soluble than polar hydrocarbons [7].

2.4. Safety Profile

As carbon content is increased, alcohol saturation pressure is decreased. When compared with ethanol in relation to vapor pressure and flash point, n-butanol possesses lower vapor pressure and higher flash point [8]. Due to this property it was found to be effective and stable at high temperature and safe in transporting.

3. Types and Isomers of Butanol

Butanol is an organic compound with the four-carbon structured alcohol moiety present at terminal positions with a molecular formula of C_4H_9OH. It is also known as butyl alcohol. Butanol composed of four isomeric structures staring from a straight primary chain alcohol to branched tertiary alcohol. Butanol have excellent solvent properties due to which it was used as fuel and intermediate in chemical synthesis. Butanol was produced from many sources such as via biological sources which acts as a precursor for fermented food and waste products. The butanol produced by these methods was to be known as bio-butanol [9]. The structural elucidation of four isomeric forms are depicted in Figure 1.

n – butanol   Iso-butanol   Secondary butanol   Tertiary butanol

Figure 1. Structural description of four Isomeric forms of Butanol.

The four isomeric forms of butanol were classified using the change in the position of carbon and alcohol group differentiate it into different isomeric forms. The butanol with a straight chain of carbons with the functional group of alcohol present at terminal position. The structured compound is known as n-butanol or primary butanol. By changing the position of an alcohol with carbon which is in internal, secondary butanol is formed. A terminal carbon possess branched isomer containing alcohol i.e., 2-methyl-1-propanol and internal carbon possess branched isomer containing alcohol i.e., tert-butanol or 2-methyl-2-propanol [10].

These all 4 butanol isomeric forms tends to have variation in melting and boiling points. The solubility is limited in n-butanol and isobutanol while secondary butanol has more solubility. At high temperature tertiary

Butanol is also miscible with water. The presence of hydroxyl group increase the polar nature due to which solubility enhances.

3.1 First Isomeric Form of Butanol -n-Butanol

n-Butanol was also known as normal butanol. It is a primary alcohol with a 4 carbon structured moiety present in it. The molecular formula of n Butanol was found to be C_4H_9OH. It was mainly produced by fermentations of sugars and carbohydrates which are mainly found in the carbohydrates present in beverages as well as food. It is widely used as artificial flavoring agent in cream and baked goods etc. It was also used as an intermediate in industry for the fabrication of solvent like butyl acetate. Propylene helps to provide petrochemical activity to n-butanol [11].

Properties of n-Butanol
The presence of a hydroxyl group (–OH) attached with one carbons atoms of the molecule differs its form other moieties of its isomeric forms.

Production
Butanol is the natural product obtained and it follows ABE fermentation method technology.

Applications

- It is widely used as a solvent in various paints, coatings, varnishes.
- It was also used as Plasticizer which was used to enhance the plastic material process.
- It is used as chemical intermediate or raw material for chemicals and Plastics.
- It is also used as swelling agent from coated fabric in textile industry.

- It was also used in various cosmetic products such as make up, nail care products and shaving products.
- It is also used in drugs such as in antibiotics, hormones and vitamins.
- It is also used in gasoline (as an additive) and brake fluid (formulation component).

3.2. Second Isomeric Form of Butanol - Secondary Butanol or 2 Butanol

Secondary Butanol also known as 2 butanol is an organic compound with a molecular formula of $CH_3CH(OH)CH_2CH_3$.

Properties

Secondary alcohol is a highly flammable, colorless liquid which is solubilized in organic solvents whereas partially soluble in water. It is produced on a larger scale using the precursor such as methyl ethyl ketones. 2-Butanol is chiral and exhibits two stereoisomerism forms such as (R)-(−)-2-butanol and (S)-(+)-2-butanol. It is used in the ratio of 1:1 mixture of the two stereoisomers in a racemic mixture [12].

Production

At Industrial Scale, Secondary butanol is produced through hydration process of primary and secondary butane with the help of sulphuric acid, which was used as a catalyst for this conversion. The process of reaction was shown in Figure 2.

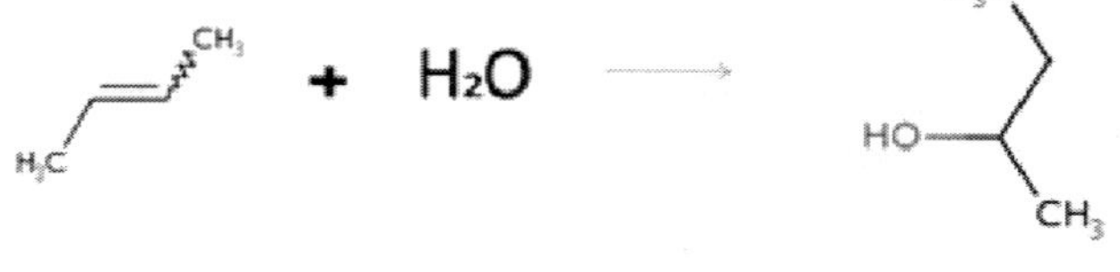

Figure 2. Chemical reaction of formation of butanol through the hydration of primary butane.

At the laboratory level, it can be prepared by using gringard reaction by using ethyl magnesium bromide with acetaldehyde or reacting with diethyl ether or tetrahydrofuran.

Applications

Butanol is highly applied as solvent because it got converted into butanone (methyl ethyl ketone). The butanone formed was being extensively used for cleaning and used for paint removals bears a pleasant aroma contains the volatile esters made their numerous applications in perfumes or in artificial flavours [16].

3.3. Third Isomeric Form of Butanol -Iso-Butanol

Isobutanol also known as 2-methylpropan-1-ol is compound possessing organic property with the molecular formula $(CH_3)_2CHCH_2OH$.

Properties

The compound is colorless, flammable liquid with characteristic smell is used as a solvent directly or in esters.

Production

Isobutanol is produced by the carboxylation of Propylene. At industrial level it was formed by using the hydroformylation method. The reaction is carried by using cobalt and rhodium complexes. The resultant compound was hydrogenated to alcohols which were then separated and isolated [13].

$$CH_3CH=CH_2 + CO + H_2 \rightarrow CH_3CH_2CH_2CHO$$

Laboratory Synthesis

Propanol and Methanol can also be used and combined to form isobutyl alcohol.

3.4 Fourth Isomeric Form of Butanol- Tert-Butanol

Tertiary butanol is the simplest form of tertiary alcohol with a molecular formula of $(CH_3)_3COH$.

Property

Tertiary butyl alcohol is a colourless solid which melts at room temperature and have a pleasant odour like camphor. It is miscible with water, ethanol and diethyl ether. Tertiary butyl alcohol mainly found in the beer and chick peas. It is rich in cassava which was used as fermenting ingredient in alcoholic beverages [14].

Production

Tertiary butyl alcohol was obtained from isobutene at commercial scale which was produced as co-product when propylene oxide is being manufactured. It is mainly formed by using gringard reaction using methyl magnesium chloride and acetone with isobutylene which undergoes catalytic hydration results in the formation of azeotrope. The large amount of water was air dried to remove majority of the solvent. On the other hand, small amount of water can be removed by treating with calcium oxide, potassium carbonate, calcium sulphate and magnesium sulphate using fractional distillation. The resulting anhydrous tertiary butyl alcohol was further obtained by refluxing and distilling from magnesium and alkali [15].

By concluding all the four isomeric forms their characteristics were briefly discussed in Table 2.

Table 2. Comparison of four Isomeric Properties of Butanol

Property	n- Butanol	IsoButanol	Secondary Butanol	Tertiary Butanol
Density at 20 degree Celsius	0.810	0.8020	0.806	0.781
Boiling point	118	108	99	82
Water solubility (g/100ml)	7.7	8.0	12.5	Miscible
Flash point (Celsius)	35	28	24	11
Auto ignition temperature (Celsius)	345	415	406	470
Explosive limits (%)	1.4-11.3	1.7-10.9	1.7-9.0	2.4-8.0

4. METHODS OF PRODUCTION OF BUTANOL

For butanol production, number of methods have been designed using different substrates and chemical methods were discussed below such as production of butanol from corns, biomass, ABE method with different microorganisms depending on the substrate and primer used, from food waste, form fruit residues, hydrolysis of plants and using various chemical methods such as dehydrogenation, Aldol condensation and hydrogenation take place. The resultant output yield percent concentration of the final product varies according to the mode of method of production used. The different mentioned methods were discussed below as follows:

4.1. Production from Biomass

In this method, the butanol was produced from bio mass using two microbes such as Clostridia and Enteric bacteria. The Clostridia are the gram-positive bacteria with rod shaped, slightly anaerobes and produce polysaccharides from the substrates such as pentose. Enteric Bacteria undergoes process the genetic manipulation along with physiological methods to produce butanol [17].

4.2. Production of Butanol from Corn

In this method the microorganisms such as Clostridia which are gram positive were used for the spore formation. The temperature required for the fermentation varies between 30-40 degree Celsius and pH is maintained at 7. Carbohydrates were used in this method which enhances the activity of cellulose even in concentration of solvents [18]. A phosphoenol pyruvate dependent transferase system is important for uptake of carbohydrates for Clostridium bacteria. A biphasic fermentation method is employed for production of butanol.

4.3. ABE Technique for Production of Butanol from Clostridia and Enteric Bacteria

In the ABE method, during metabolic pathway two phases were used such as acidogenesis and solventogenesis were used for the production of Butanol. Acidogenesis and solventogenesis involves following sequence of steps such as followed by starting initially with fermentation. In the abiogenesis method the substrate get converted into pyruvate followed by acidogenesis get converted into acetate and butyrate [19]. In the second phase solventogenesis process takes place in which acetate was formed from acetone and butyrate leads to formation of Butanol and ethanol. Further, metabolic pathway of butanol production follows a steps such as

1. Glucose conversion to pyruvate with 2 ATP and NADPH production.
2. Pyruvate gets converted in to acetyl-Co-A using enzyme pyruvate-ferrodoxinoxidoreductase.
3. Acetyl-Co-A converted in to Aceto-acetyl-Co-A through acetyl transferase enzyme through 3-hydroxyl-CoA dehydrogenase enzyme [20].
4. Further conversion of Aceto-acetyl-Co-A in to 3-hydroxybutyryl-Co-A through 3-hydroxyl-CoA dehydrogenase enzyme.
5. 3-hydroxybutyryl-Co-A converted into Crotceryl-Co-A by Crotonase enzyme.
6. Crotceryl-Co-A converted into Butyryl-Co-A by Butyryl CoA dehydrogenase enzyme.
7. Moving further, Butyryl-Co-A converted into Butyraldehyde through Butyraldehyde dehydrogenase enzyme [21].
8. Finally, Butanol is formed from Butyraldehyde by Butanol Dehydrogenase enzyme.

Various Process for Production of Butanol

a. *Batch Process-* The simplest and most common method employed for butanol fermentation as it is free from any contamination and all the process undergone is not complicated one [22]. There are certain difficulties in this operation such as the batch produced from reactor implements less yield from the mentioned one in the lag phase along with manufacturing steps like manufacturing, sterilizing and filling time is not maintained during batch reactor process. In this process, glucose take part as an substrate for production of 33 g/L butanolin fermentation which takes a time of 72 hours [23].

b. *Fed-Batch Fermentation-* This process includes mode of batch in reactor process with have less amount of volume of media along with concentration of substrate. As the reaction time is accelerated, the initial volume of culture accelerates during this process. The research states that toxicity of butanol is pretty high against C.acetobutylicum, A new novel method should be employed for final estimated yield separation. Due to the low amount of substrate in final phase, the growth of cell is accelerated which ultimately results in high reactor utility. The advantage of this method is that it involves prevention of substrate accumulation, produces high cell amount.

c. *Continuous-Culture Technique-* This process leads to exponential phase in which there is tremendous cell growth as batch mode is being made for reactor process. During the whole process, continuous transfer of media in reactor is mandatory and less time is required for product to be released to maintain healthy amount of feed in the reactor. In this process, 1.74g/l butanol is produced from C. Beijerinckii in a good batch process.

d. *Immobilized Cell Reactor-* This reactor implements a reactor process which tends to produce healthy amount of Butanol i.e., 50 times than previous process. It carries an advantage that cell concentration is always maintained in the reactor and it is calculated more than the previous methods.

e. *Membrane Cell Reactor*-It involves ultrafiltration process which involves fibre shield with hollow membrane that ultimately removes cells in the continuous mode. In this process, attached hollow membrane may break thus immobilized system is employed for filtration process [24].

4.4. Production of Butanol from Fruit Residues

This process involves Clostridium acetobutylicum which was used as microorganism in the ABE fermentation method. The fermented products were placed in fermentation bottles containing medium which is synthetic in nature and contain glucose extract undergo anaerobic conditions. The main medium includes apples as well peers obtained in bulk quantity whose peels are removed by knife which are in 1-2cm in appearance. They are dried in air and diluted with distilled water which undergoes extraction process. After extraction, the peels undergoes treatment with sugar solution. Further, method of batch fermentation is employed which involve bottle made of pyrex carrying media amount of 70 ml placed in fermenters. The temperature is set at 37 degree Celsius. The inoculation of media is done with 6.25% (v/v) suspension and every sugar as well as cell characterization is done with 3ml cultures. Analytical procedure is being carried out which employ pH measurements and sugar contents measurements done by HPLC mode. Internal standard method maintain butanol amount. Finally, all the results of fermentation method is being obtained by amount of sugar present in apple peels extract which proves that that C. acetobutylicum is best agricultural industrial main residue to produce Butanol [25].

4.5. Production of Butanol from Food Wastes

It involves media such as Clostridium medium devoid of glucose content dissolved in whey food wastes that contain lactose and sterilization is done at 120 degree Celsius for 20 minutes. Inoculation of whey in medium is done after 24 hours. A spore culture is formed containing 50ml batch cultures and gas chromatography method was employed for measuring the concentration of solvents. Further, fuels are mixed with butanol prepared on calculation such as basis of weight. The final butanol is formed in the range of 0-20% w/w which is detected by direct fuel injection system. [26].

4.6. Production of Butanol by Hydrolysis of Plants

This process involves production of butanol in a very simple manner which involves fermentation from hexose sugars. These sugars are mostly obtained from process of hydrolysis of plants that are rich in starch i.e., wheat and rice. Initially, all the key ingredients are converted in to dextrose by hydrolysis process. Ultimately, dextrose content accumulated in high concentration which gets converted in to glucose by glucoamylase enzyme. This produces high yields of butanol as proved by the persons undergoing research on butanol production. Research found that glucose is the main origin for producing 014gl of butanol. This undergoes a fermentation process which should be continuous in nature and must be in closed circuit. pH must be maintained at 4. The glucose content for butanol production must be maintained around 30gL in the exceeding range of fermentation which have no effect on Clostridium acetobutylicum strain [27].

4.7. Chemical Methods

This involves production of butanol from n-ethanol by three methods such as

a. Initially, a phase in which liquid reaction take place which produce dehydrogenation of ethanol to form aldehyde
b. Secondly, Aldol condensation of aldehyde take place
c. Finally, butanol hydrogenation takes place.

It is a gas phase reaction of ethanol which doesn't produce intermediates like acetaldehyde and include zeolite catalyst to form butanol. This whole mechanism carried in industry which helps to increase alcohol carbon number. Various catalysts such as magnesium and aluminum mixed oxides, nickel and aluminum oxide and magnesium oxide can be used in this reaction [27, 28].

5. Strategies for Enhancement of Butanol Production

There are various enhancement strategies for production of butanol in various cases such as

1. In fermentation process of Clostridium, the yield of butanol increased to 17g/L with the presence of native bacteria. Various scientific researches proved this through correct specific amount of bacterial strain, carbon origin which must be of low cost, process variables and various metabolic and recovery management. The processing is improved through immobilization of cell process, biomass recycling and lowering content of inhibitor in fermentation process [29].
2. In case of improvement of process, use of expensive glucose can be replaced by different cases of hydrolysis of straw wheat in fermentation process of butanol. Various combination process used for batch processing of butanol which involves steps like wheat straw previously treated fermentation, mixing as well as hydrolysis of wheat straw and removal of sediments from wheat straw [30].
3. Further inhibition of product and substrate, salt concentration, presence of dead cells, lack of oxygen and nutrient addition in fermenter affect butanol fermentation.

4. If lime is reacted with straw hydrolyses, than production of butanol increases. Moreover, conversion of lignin biomass to butanol also helps to increase the production.
5. Various sterilization as well as manufacturing procedures improve the working as well as efficiency of batch and fed fermentation method of butanol production [31].
6. If the case of continuous fermentation processes, renewal cell lining as well system, immobilized cells and various agitation process improve the microbial along with nutrient content, helps in large quantity cell transfer.
7. Various process like immobilization helps in lag phase removal that helps in updating the process and producing reliable operation eliminating continuous inoculation during course of process.
8. The process variables such as different biocatalyst use and cell stability also helpful in productivity of butanol in large content.
9. Depending on the type of bioreactor such as fibrous bed bioreactor, from corn strain a high yield of butanol is produced in continuous fermentation process.
10. Acidogenesis process can be made more productive by use of co-substrate of butyric acid in stream of feed [32].
11. Solventogenesis process can be made productive by increasing the validity of various solvent systems in stream of feed.
12. In continuous fermentation method, recycling of cells is improved by filtration panel presence in the bioreactor which increases efficiency of process.
13. Various distillation process that provide distillation column leads to purification of butanol in high amount [33].
14. New innovation in fermentation techniques such as flash fermentation increase the efficiency and productivity of butanol to good extend.
15. The removal of left over butanol is done by extraction method such as liquid-liquid extraction, osmosis and adsorption process.
16. Metabolic engineering process helpful in removal of destruction cells in bacterial strains which is helpful in production of butanol

17. If there is random process of mutagenesis to metamorphose, then changes in genes DNA sequence leads to high proportion of butanol
18. If *C. acetobutylicum* treated undergoes exposure of UV rays than butanol production is increased by 20% in molasses strain.
19. New pathway such biosynthetic production of butanol is possible by use of various metabolic engineering system levels of clostridia.
20. In genetic engineering process, recombinant DNA technology provides healthy production of solvent for high efficiency of butanol production [34].

6. APPLICATIONS OF BUTANOL IN VARIOUS FIELDS

6.1. Butanol as a Biofuel

Bioethanol was produced on industrial scale in the form of biomethane. Biobutanol is an upcoming fuel supplement for gasoline, diesel and kerosene. It was mainly produced by biological derivatives which are used to produce alcohol. This bio based alcohol was used as a solvent in various chemical industries. Butanol is also used a renewable biofuel which was used in internal combustion engines possess its great advantage over gasoline, diesel fuel and biofuels such as methanol, ethanol, biodiesel. The n-Butanol also possess advantages over the drawbacks of low carbon alcohols or biodiesels [35].

6.2. Industrial Use of Butanol

Butanol is widely used as solvent in industry such as textile as well as chemical on a vast scale. It was also used various chemical and organic synthesis. It acts as thinner and solvent for coating applications. It is also used as brake and hydraulic fluid component. Synthesis of 2-butoxyethanol is applicable through Butanol. The main application of butanol was found that it acts as reactant for production of butyl acrylate from acrylic acid which have vast application in acrylic paint containing water. Due to high aromatic profile, it is highly suited as perfume base [36].

6.3. Uses of Butanol in Cosmetic Industry

1. Used as humectant for cellulose nitrate.
2. Used in the cosmetic industry as well in products such as shampoo, shaving products and soaps.
3. As a chemical intermediate to create other vital compounds such as Glycol Ether, Acrylate Esters, Amino Resins, Acetates, and Amines.
4. Used as solvent for paints, coatings, varnishes
5. Used as plasticizers which were used to enhance the plastic material potential.
6. Used as coating agent in numerous applications.
7. Used as chemical intermediate or raw material for the production of various chemicals and raw materials.
8. Used in various drugs, antibiotics, hormones and vitamins.
9. Used as a Gasoline (as an additive) and brake fluid (formulation component)
10. Used as dispersing agent to form clear preparations and floor polishes.
11. Used in manufacture and processing of extracts for pharmaceutical goods and Pesticidal formation [35,36]

Main Problems

Butanol production was still challenging task due to the rapid extinction of existing fuels creates a stress bearing pressure on production of biobutanol to meet the demands of upcoming era and global health. The work with natural butanol producers is very challenging. The species used for fermentation such as Clostridia are not much suitable and appropriate as they have developed intolerance to oxygen as a result of which results in Butanol toxicity. Moreover, butanol is not synthesized in economically in desirable quantities. Therefore a need to rise and enhances the production of butanol by using appropriate microbes have been risen. Various genetic engineering methods have been used for the manipulations of genes results in acid formation and solvents accumulations [37].

CONCLUSION AND FUTURE PROSPECTIVE OF BUTANOL

Much debates and discussions have been carried out on the future prospects of Butanol. As we have seen in the former part, Butanol was sued a commercial biofuel which have been replaced gasoline and diesel with wide variety of industrial applications, in pharmaceuticals such a solvent organic compound, in cosmetic industry, textile industry. Moreover, the recent approach has been surrounded by various issues such as cost effectiveness, safety, health hazards, environmental safety and climate safety and toxicity etc. and their effects on the global economics were speculated. In spite of the existing disputes, the Butanol as a biofuel renders massive hope for the betterment of future. The innovations in all the aspects and all the fields can be made with its principles. Miles we have walked in this field and miles must go. The technology is showing up hopes for the future even after some disputes. The use of this Butanol production via different modes of inexpensive renewable non-food carbon substrates would solve and eradicate the crisis of world's problems such as like global warming and the over-consumption of petroleum based products.

REFERENCES

[1] Grodowska, Katarzyna, and Andrzej Parczewski. "Organic solvents in the pharmaceutical industry."*Acta Pol Pharm*67, no. 1 (2010): 3-12.
[2] Slater, C. S., M. J. Savelski, R. P. Hesketh, and E. Frey. "The selection and reduction of organic solvents in pharmaceutical manufacture." In

Washington: 10th Green Chemistry and Engineering Conference. 2006.

[3] Afschar, A. S., H. Biebl, K. Schaller, and K. Schügerl. "Production of acetone and butanol by Clostridium acetobutylicum in continuous culture with cell recycle." *Applied microbiology and biotechnology* 22, no. 6 (1985): 394-398.

[4] Sitaramaraju, Yarramraju, AdilRiadi, Ward D'Autry, Kris Wolfs, Jos Hoogmartens, Ann Van Schepdael, and Erwin Adams. "Evaluation of the European Pharmacopoeia method for control of residual solvents in some antibiotics." *Journal of pharmaceutical and biomedical analysis* 48, no. 1 (2008): 113-119.

[5] Makitra, R. G. "Reichardt, C., Solvents and Solvent Effects in Organic Chemistry, Weinheim: Wiley-VCH, 2003, 630 p." *Russian journal of general chemistry* 4, no. 75 (2005): 664-664.

[6] Lebedevas, Sergejus, Galina Lebedeva, EgleSendzikiene, and VioletaMakareviciene. "Investigation of the performance and emission characteristics of biodiesel fuel containing butanol under the conditions of diesel engine operation." *Energy & fuels* 24, no. 8 (2010): 4503-450

[7] Wu, Han, KarthikNithyanandan, Timothy H. Lee, Chia-fon F. Lee, and Chunhua Zhang. "Spray and combustion characteristics of neat acetone-butanol-ethanol, n-butanol, and diesel in a constant volume chamber." *Energy & fuels* 28, no. 10 (2014): 6380-6391.

[8] Zheng, Zun-Qing, Shan-Ju Li, Hai-Feng Liu, Ming-Fa Yao, JiaXu, and Bin-Bin Yang. "Mechanism of effects of n-butanol properties on low temperature combustion in a diesel engine." *Transactions of Chinese Society for Internal Combustion Engines* 31, no. 2 (2013): 97-102.

[9] Moss, Jeffrey T., Andrew M. Berkowitz, Matthew A. Oehlschlaeger, Joffrey Biet, Valérie Warth, Pierre-Alexandre Glaude, and Frédérique Battin-Leclerc. "An experimental and kinetic modeling study of the oxidation of the four isomers of butanol." *The Journal of Physical Chemistry*A112, no. 43 (2008): 10843-10855.

[10] Sarathy, S. Mani, StijnVranckx, Kenji Yasunaga, Marco Mehl, Patrick Oßwald, Wayne K. Metcalfe, Charles K. Westbrook et al. "A comprehensive chemical kinetic combustion model for the four butanol isomers." *Combustion and flame*159, no. 6 (2012): 2028-2055.

[11] Van Geem, Kevin M., Steven P. Pyl, Guy B. Marin, Michael R. Harper, and William H. Green. "Accurate high-temperature reaction networks for alternative fuels: butanol isomers." *Industrial & engineering chemistry research* 49, no. 21 (2010): 10399-10420.

[12] da Silva Trindade, Wagner Roberto, and Rogério Gonçalves dos Santos. "Review on the characteristics of butanol, its production and use as fuel in internal combustion engines." *Renewable and Sustainable Energy Reviews* 69 (2017): 642-651.

[13] Ansón, Alejandro, Rosa Garriga, Santiago Martínez, Pascual Pérez, and Mariano Gracia. "Densities and viscosities of binary mixtures of 1-chlorobutane with butanol isomers at several temperatures." *Journal of Chemical & Engineering Data*50, no. 2 (2005): 677-682.

[14] Chmielewska, Agnieszka, Małgorzata Żurada, Krzyszt of Klimaszewski, and Adam Bald. "Dielectric properties of methanol mixtures with ethanol, isomers of propanol, and butanol." *Journal of Chemical & Engineering Data* 54, no. 3 (2009): 801-806.

[15] Xu, Yihui Tom, and William D. Parten. "*Recovery of butanol isomers from a mixture of butanol isomers, water, and an organic extractant.*" U.S. Patent 8, 968,522, issued March 3, 2015.

[16] Wang, Yan, Tai Shung Chung, Huan Wang, and Suat Hong Goh. "Butanol isomer separation using polyamide–imide/CD mixed matrix membranes via pervaporation." *Chemical engineering science* 64, no. 24 (2009): 5198-5209.

[17] Zhao, Jingbo, Congcong Lu, Chih-Chin Chen, and Shang-Tian Yang. "Biological production of butanol and higher alcohols." *Bioprocessing technologies in biorefinery for sustainable production of fuels, chemicals, and polymers* (2013): 235-262.

[18] Nagarajan, Vasantha, and Michael G. Bramucci. "*Yeast production culture for the production of butanol.*" U.S. Patent Application 13/162,165, filed June 14, 2012.
[19] Green, Edward M. "Fermentative production of butanol—the industrial perspective." *Current opinion in biotechnology* 22, no. 3 (2011): 337-343.
[20] Karimi, Keikhosro, Meisam Tabatabaei, IlonaSárvári Horváth, and Rajeev Kumar. "Recent trends in acetone, butanol, and ethanol (ABE) production." *Biofuel Research Journal* 2, no. 4 (2015): 301-308.
[21] Abdehagh, Niloofar, F. Handan Tezel, and Jules Thibault. "Separation techniques in butanol production: challenges and developments." *Biomass and Bioenergy* 60 (2014): 222-246.
[22] Amiri, Hamid, and Keikhosro Karimi. "Improvement of acetone, butanol, and ethanol production from woody biomass using organosolv pretreatment." *Bioprocess and biosystems engineering* 38, no. 10 (2015): 1959-1972.
[23] Andersch, Wolfram, Hubert Bahl, and Gerhard Gottschalk. "Level of enzymes involved in acetate, butyrate, acetone and butanol formation by Clostridium acetobutylicum." *European journal of applied microbiology and biotechnology* 18, no. 6 (1983): 327-332.
[24] Dürre, Peter. "Biobutanol: an attractive biofuel." *Biotechnology Journal: Healthcare Nutrition Technology* 2, no. 12 (2007): 1525-1534.
[25] Bennett, George N., and Frederick B. Rudolph. "The central metabolic pathway from acetyl-CoA to butyryl-CoA in Clostridium acetobutylicum." *FEMS microbiology reviews* 17, no. 3 (1995): 241-249.
[26] Dürre, Peter. "New insights and novel developments in clostridial acetone/butanol/isopropanol fermentation." *Applied microbiology and biotechnology* 49, no. 6 (1998): 639-648.
[27] Ezeji, Thaddeus Chukwuemeka, Nasib Qureshi, and Hans Peter Blaschek. "Bioproduction of butanol from biomass: from genes to bioreactors." *Current opinion in biotechnology* 18, no. 3 (2007): 220-227.

[28] Malaviya, Alok, Yu-Sin Jang, and Sang Yup Lee. "Continuous butanol production with reduced by products formation from glycerol by a hyper producing mutant of Clostridium pasteurianum." *Applied microbiology and biotechnology* 93, no. 4 (2012): 1485-1494.

[29] Burlew, Keith H., Michael Charles Grady, John W. Hallam, David J. Lowe, Brian Michael Roesch, and Joseph J. Zaher. "*Methods and systems for removing undissolved solids prior to extractive fermentation in the production of butanol.*" U.S. Patent 8,557,540, issued October 15, 2013.

[30] Zheng, Yan-Ning, Liang-Zhi Li, Mo Xian, Yu-Jiu Ma, Jian-Ming Yang, XinXu, and Dong-Zhi He. "Problems with the microbial production of butanol." *Journal of industrial microbiology & biotechnology* 36, no. 9 (2009): 1127-1138.

[31] Lee, Sang Yup, Jin Hwan Park, SehHee Jang, Lars K. Nielsen, Jaehyun Kim, and Kwang S. Jung. "Fermentative butanol production by Clostridia." *Biotechnology and bioengineering* 101, no. 2 (2008): 209-228.

[32] Jin, Chao, Mingfa Yao, Haifeng Liu, F. Lee Chia-fon, and Jing Ji. "Progress in the production and application of n-butanol as a biofuel." *Renewable and sustainable energy reviews* 15, no. 8 (2011): 4080-4106.

[33] Nielsen, David R., Effendi Leonard, Sang-Hwal Yoon, Hsien-Chung Tseng, Clara Yuan, and Kristala L. Jones Prather. "Engineering alternative butanol production platforms in heterologous bacteria." *Metabolic engineering* 11, no. 4-5 (2009): 262-273.

[34] Xue, Chuang, Jingbo Zhao, Lijie Chen, Shang-Tian Yang, and Fengwu Bai. "Recent advances and state-of-the-art strategies in strain and process engineering for biobutanol production by Clostridium acetobutylicum." *Biotechnology advances* 35, no. 2 (2017): 310-322.

[35] Ndaba, Busiswa, Idan Chiyanzu, and Sanette Marx. "n-Butanol derived from biochemical and chemical routes: A review." *Biotechnology Reports* 8 (2015): 1-9.

[36] Terabe, Shigeru. "Capillary separation: micellarelectrokinetic chromatography." *Annual Review of Analytical Chemistry* 2 (2009): 99-120.

[37] Garti, N., A. Aserin, and M. Fanun. "Non-ionic sucrose esters microemulsions for food applications. Part 1. Water solubilization." *Colloids and Surfaces A: Physicochemical and Engineering Aspects* 164, no. 1 (2000): 27-38.

In: Properties and Uses of Butanol
Editor: Arnaud M. Artois

ISBN: 978-1-53618-448-8
© 2020 Nova Science Publishers, Inc.

Chapter 2

PRODUCTION OF BUTANOL FROM BIOMASS: ADVANCES AND CHALLENGES

Fernando Israel Gómez-Castro[1],,
Eduardo Sánchez-Ramírez[1],
Ricardo Morales-Rodriguez[1],
Juan José Quiroz-Ramírez[2]
and Juan Gabriel Segovia-Hernández[1]*

[1]Departamento de Ingeniería Química, Universidad de Guanajuato, Guanajuato, Guanajuato, Mexico
[2]Centro en Innovación Aplicada en Tecnologías Competitivas (CIATEC A.C.), León, Guanajuato, Mexico

ABSTRACT

Butanol is an alcohol that can be used as fuel in internal combustion engines as a replacement for gasoline. Even though ethanol is the most known alcohol for such use, it is commonly employed as an additive

---

* Corresponding Author's Email: fgomez@ugto.mx.

instead of a fuel, since it cannot be used 100% in the existing engines. On the other hand, butanol has similar properties to those of gasoline, turning it in a better alternative to run vehicle engines. Like ethanol, butanol can be produced from sugar/starchy biomass and lignocellulosic biomass. Butanol produced from such raw materials is known as biobutanol. The traditional production process for that product is known as ABE fermentation since acetone, butanol, and ethanol are obtained. When lignocellulosic materials are used as raw material, pre-treatments are required before the fermentation step. Although ABE fermentation is a relatively known process, there are still several challenges to turn biobutanol into an economically competitive product. In this chapter, the main aspects involved in the production of biobutanol are described, including raw materials, the transformation of biomass and the separation of the ABE mixture. The most important areas of opportunity are described, focusing on the enhancements required by the production process to increase reaction yields in the hydrolysis and fermentation steps; and to reduce costs and reduce environmental impact in the purification trains.

Keywords: bio-butanol, ABE fermentation, purification technologies

INTRODUCTION

Crude oil is the most important source of energy accounting for 35% of the world's energy consumption. With the growing demand for crude oil, experts have estimated that the "peak oil," where half of the worldwide resources are depleted and the delivery rates cannot be satisfied, was met in 2010 (Dautidis et al., 2013). Although crude oil, natural gas, and coal will still remain the most important sources of energy until at least 2030, the problem of depleting oil reserves has been recognized as well as the fact that using renewable resources leads to a higher security of supply, a better environment, a higher national added value and an increase in income in rural regions. Biofuels are being promoted on a global basis because their use may potentially reduce greenhouse gas emissions and help achieving energy security by reducing the transportation sector's use of and reliance upon petroleum. Renewable raw materials are receiving increasing interest in the context of next-generation liquid transportation fuels (Hechinger et al., 2010). Today's fossil fuels mostly consist of long-chained hydrocarbon mixtures; where they can be efficiently produced from crude oil while offering high volumetric energy content. Combustion engines have been designed to operate on these blends; consequently, current research largely focuses on the development of bio-based processes that aim at next-generation fuels that mimic the fuel structures currently produced from fossil feedstock. However, instead of imitating the established fossil value chains, new products and processes should be investigated which reflect the tremendous influence of the raw material change from fossil to biomass feedstock. Biofuels derived from low-input high-diversity mixtures of native grassland perennials can provide more usable energy, greater greenhouse gas reductions, and less agrochemical pollution per hectare than corn grain ethanol or soybean biodiesel (Tilman et al., 2006). High-diversity grasslands had increasingly higher bioenergy yields that were 238% greater than monoculture yields after a decade. Biofuels derived from low-input high-diversity are carbon negative because net ecosystem carbon dioxide sequestration (4.4 megagram hectare−1 year−1 of carbon dioxide in soil and roots) exceeds fossil carbon dioxide release during biofuel production (0.32 megagram hectare−1 year−1). Moreover, biofuels can be produced on agriculturally degraded lands and thus need to neither displace food production nor cause loss of biodiversity via habitat destruction (Tilman et al., 2006).

Biomass conversion to fuels and chemicals can, therefore, play a key role in mitigating the heavy dependence on fossil carbon in the chemical and petrochemical industries. Recognizing this, legislation and policies have been formulated worldwide to mandate and incentivize biomass conversion to fuels. For example, renewable fuel standards (the US federal mandates on volumes for blending biofuel into transportation fuel) require the production of over 36 billion gallons of biofuels, including over 16 billion gallons of cellulosic and other advanced biofuels by 2022 (Hechinger et al., 2010; Dautidis et al., 2013). Incentives such as the biomass crop assistance program provide financial assistance to farmers and operators to grow and deliver biomass for advanced biofuel production. Mandates for producing biofuels and incentives for installing high biofuel blend pumps have come into effect in several states. Similar directives and policies toward greater biofuel production have also been formulated in the EU (biofuels directive 2003/30), China, and India (Wang et al., 2011a; Dautidis et al., 2013). Several technical challenges exist in developing biomass-based fuels and the chemical industry. Biomass has a C:O ratio up to 1:1 (for carbohydrates); it is, therefore, significantly different in composition from crude oil. Furthermore, oxygen is present in different functional forms, such as hydroxyl, ether and carbonyl groups, which require different chemical steps for its removal. Numerous chemical processing steps have been developed to this end. As a result of resource availability, scientific and technological advancements, and favorable policy, the concept of "biorefinery" the biomass-based parallel of the traditional petroleum refinery, has emerged. It is envisaged that a biorefinery will draw a complex feedstock of biomass and convert it, in a series of processing steps, into valuable products (Wang et al., 2011a).

Among the different biofuels, biobutanol has been proposed as a potential replacement for gasoline. Biobutanol is butanol obtained from renewable sources. The most common butanol isomer obtained by fermentation is n-butanol (Moreno and Cubillos Lobo, 2017). Biobutanol has some advantages over bioethanol to be used as fuel in engines. Butanol has a higher heating value than ethanol, thus the fuel consumption can be reduced. Butanol has lower volatility, reducing issues as cavitation and vapor lock. Butanol has lower vapor pressure and higher flash point, turning it into a safer fuel, and it is less corrosive than butanol (Jin et al., 2011). Additionally, it has a lower affinity for water and can be mixed with gasoline in any proportion (García et al., 2011). Table 1 shows the main physical properties of n-butanol, ethanol, and gasoline.

Due to the advantages of biobutanol over bioethanol, the production of n-butanol from renewable sources has re-gained interest in the last years. In this chapter, the key factors in biobutanol production will be discussed. First, the variety of raw materials from which biobutanol can be obtained are presented. Then, pretreatment technologies are introduced. Such pretreatments are necessary when working with raw materials having undesired components difficulting the access to the sugars from which n-butanol is produced. In the following section, the fermentation procedure, known as ABE fermentation, is presented. Then, the technologies which have been developed to purify the products from the fermentation step are discussed. Next, the advances in the production of biobutanol in industrial scale are shown. Finally, the areas of opportunity in the production of biobutanol are analyzed, focusing on the further developments required to turn the renewable butanol production into economically feasible.

Table 1. Physical properties for n-butanol, ethanol, and gasoline (Balat et al., 2008; Jin et al., 2011)

	Ethanol	n-butanol	Gasoline
Self-ignition temperature (°C)	434	385	~300
Flash point (°C)	8	35	-45 to -38
Lower heating value (MJ·kg^{-1})	26.8	33.1	42.7
Latent heat at 25°C (kJ·kg^{-1})	904	582	380-500
Cetane number	8	25	0-10
Octane number	108	96	80-99
Density at 20°C (g·mL^{-1})	0.79	0.81	0.72-0.78
Viscosity at 40°C (mm^2·s^{-1})	1.08	2.63	0.4-0.8*

* at 20°C.

RAW MATERIALS

Like bioethanol, biobutanol can be obtained from mainly two kinds of raw materials: sugar/starchy materials and lignocellulosic materials. In the case of sugar/starchy materials, the conversion process is more direct, since it possible to directly access the sugar or starchy to ferment it into alcohols. In the case of lignocellulosic materials, some pretreatments are required to release the sugars and then ferment them into alcohols.

Among the sugar/starchy raw materials, crops as sugar beet (Drapcho et al., 2008), sugarcane, corn, sorghum, wheat, rye (Siegmeier et al., 2019) and barley (Gibreel et al., 2009) can be mentioned. Such crops have a high production worldwide, which made them candidates for their transformation into alcohols and other products. Sugarcane is among the crops with higher production. By 2012, 1.8 billion tons of sugarcane were produced (Silalertruksa and Gheewala, 2018). On the other hand, by 2014 around 0.27 billion tons of sugar beet have been reported (Marzo et al., 2019). In the case of corn, for 2016-2017, an approximate global production of 1.04 billion metric tons has been reported (Mohanty and Swain, 2019). For wheat, an approximate global production of 755 million metric tons has been reported in the same period (Mohanty and Swain, 2019). For barley, 141 million tons were produced by 2016 (Schnurbusch, 2019). A sorghum production of 56 million tons has been reported by 2009 (Khalil et al., 2015). Table 2 shows the sugar/starch content of these materials.

Table 2. Sugar/starch content of raw materials for the production of alcohols (Drapcho et al., 2008)

	Starch percent
Corn	65-76
Wheat	66-82
Barley	55-74
Sorghum	68-80
	Sugar percent
Sugarcane	12-17
Sugar beet	16-18

The main advantage of sugar/starchy raw materials is, as aforementioned, the relative easiness to access the sugars or starch, implying that the number of conversion stages is relatively low. Nevertheless, the use of those materials to produce biofuels has an ethic inconvenience, since they are employed as human food. Chakravorty et al. (2009) mentioned that the production of biofuels from such materials would cause increases in food prices. On the other hand, Timilsina and Shreshta (2011) indicated that such increases will be due to other factors, different from the production of biofuels. In any case, the scenario of a food shortage represents a risk that must not be taken. Thus, governments have developed policies to avoid such a situation, as Mexico, where the use of corn and sugarcane for the production of biofuels is only allowed if a surplus exists (Diario Oficial, 2008). Additionally, the use of second-generation raw materials has been promoted on the last years, to avoid the competence between the production of biofuels and the food sector. For the production of alcohols as ethanol and butanol, the second-generation materials comprise mainly agro-industrial residues, usually known as lignocellulosic materials.

Lignocellulosic materials are composed mainly of lignin, hemicellulose and cellulose. Their structure is more complex than the sugar/starchy materials, and require additional processing steps to first break the lignin, allowing the access to hemicellulose and cellulose. Among the lignocellulosic materials which can be used to produce biobutanol, sugarcane bagasse, or residues from crops as corn, sorghum, wheat, barley, among others. According to Chuck (2016), for each ton of sugarcane, 0.3 ton of bagasse are produced. In the case of corn, sorghum, wheat and barley, the residue/crop ratio is 2.0, 1.07, 1.5 and 2.25. This data is useful to estimate the availability of residues that can be used to produce biofuels and other products. As an example, according to the data reported by Mohanty and Swain (2019), about 2.08 billion metric tons of residues from corn cultivation where produced. Residues of other local crops can also be used for the production of biobutanol. As example, *Tequilana weber* is a common plant in the west of Mexico used for the production of tequila, and its bagasse has been studied for the production of bioethanol (Velázquez-Valadez, 2016; Rios-González, 2017), although its composition with around 42% of cellulose (Saucedo-Luna et al. 2010) may allow producing also biobutanol.

Composition of agro-industrial residues is variable, nevertheless, they have as common components ligning, hemicellulose and cellulose. Table 3 shows the composition of some common lignocellulosic materials, most of them with high worldwide production. Sugarcane bagasse, corn stover and barley straw show high contents of cellulose, nevertheless, sugarcane bagasse has also a high content of lignin. On the other hand, corn cobs have the highest content of cellulose, with relatively low quantities of lignin.

Other raw materials with the potential to produce alcohols have been reported, as coffee residue wastes, waste from weekly markets, and residues of confectionery production (Brosowski et al., 2016), among others. According to Choi et al. (2012) coffee production by 2010/2011 was around 8.2 million tons, and the wastes obtained in coffee processing consist of 37-42% of fermentable sugars. Organic wastes from markets are also an important source for sugars, mainly the residues of fruits. To exemplify the potential of such a source, it has been reported that the content of cellulose in orange peel ranges from 9.19 to 11.93%, while the orange bagasse has a cellulose percentage between 14.46 and 26.45% (Sánchez Orozco et al., 2014). Tiwari et al. (2014) reported the production of ethanol from rotten fruits, as apple, banana, among others. In the case of wastes from the confectionery industry, such residues have a high content of sugars. Miah et al. (2018) reported that a 4.1% of sugar-based products and 2% of chocolate products typically are left as part of the waste stream. Additionally, after packing, the other 5.7% of the products are considered as waste (Harrison et al., 2019). The high content of sugars on those wastes makes them excellent candidates for its further conversion into alcohols as bioethanol and biobutanol. The use of microalgal biomass to produce biobutanol has also been reported (Wang et al., 2017). Figure 1 summarizes the raw materials with potential for biobutanol production.

Table 3. Mean composition of common lignocellulosic materials (Castro-Montoya and Jiménez-Gutiérrez, 2013; García-Aparicio et al., 2007)

	wt% on dry basis		
	Cellulose	Hemicellulose	Lignin
Sugarcane bagasse	39.01	24.91	23.09
Corn stover	34.61	22.21	17.69
Corn cobs	45	35	15
Sweet sorghum	22.48	13.81	11.34
Wheat straw	16.85	32.64	22.63
Barley straw	37.50	25.00	11.50

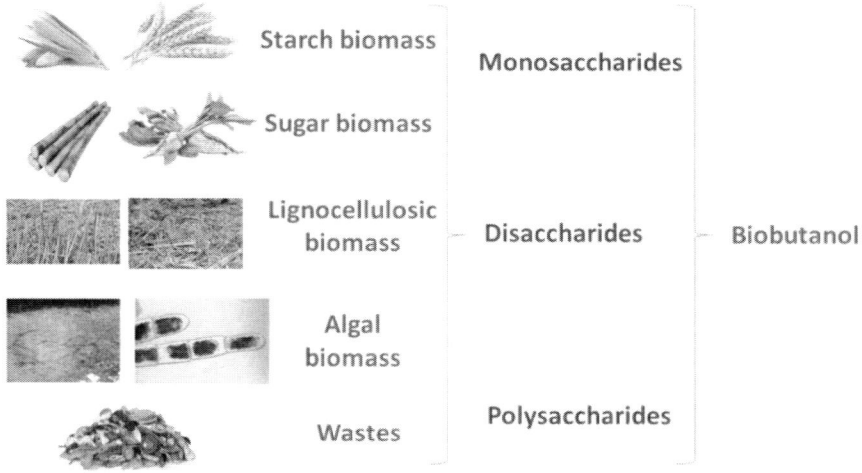

Figure 1. Raw materials with potential for the production of biobutanol.

PRETREATMENT TECHNOLOGIES

The pretreatment of biomass raw material is a crucial and necessary step in the bioproducts production. The main objective of the pretreatment is the breakdown of the lignocellulosic matrix aiming to remove lignin, maintaining the hemicellulose, reducing cellulose crystallinity, and increasing the porosity of the material (Chiaramonti et al., 2012). There are various types of pretreatment assessed to that end. Figure 2 shows the classification of the different types of pretreatments employed in the lignocellulosic biomass, as well as the main characteristic when it is implemented. In general, the pretreatment technologies can be organized in four categories: physical, chemical, physicochemical, and biological. Of course, there are some drawbacks that characterize each of the different methods, such as high cost, energy consumption, low efficiency, and specially the generation of toxic compound or inhibitors for the microorganisms used to produced the desired products, in this case ABE.

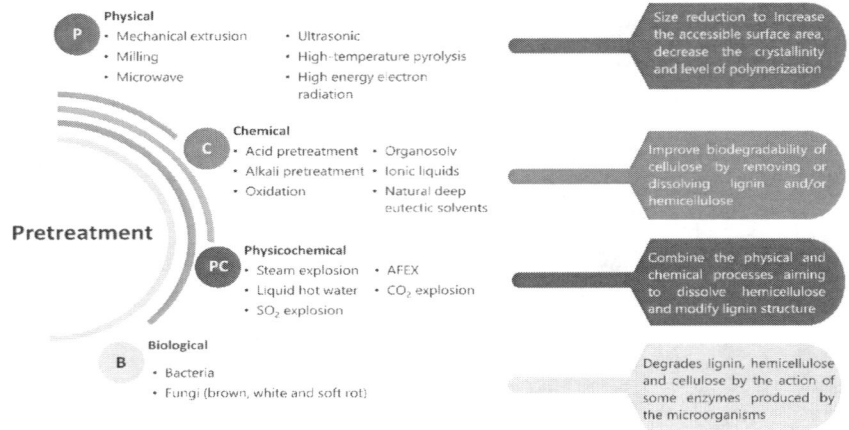

Figure 2. Classification of the different types of pretreatment methods employed in the lignocellulosic biomass (Behera et al., 2014; Chen et al., 2017; Chiaramonti et al., 2012).

PHYSICAL PRETREATMENT

The physical pretreatment is employed to increase the accessible area and pore size of lignocelluloses, as well as decreasing the crystallinity and the degree of polymerization. The most common pretreatments are mechanical extrusion, milling, microwave, ultrasonic, high-temperature pyrolysis, and high energy electro radiation. Table 4 shows the main characteristics of the physical pretretaments.

Table 4. Physical pretreatment and main characteristics (Chen et al., 2017)

Type of pretreatment	Main characteristics
Mechanical extrusion/ splintered	Common methods: dry crushing, wet crushing, vibrating ball mill grinding, and compression. The particle size of the material can be reduced to increase the contact surface. Size reduction is one of the most effective methods to increase the enzymatic accessibility to lignocelluloses. Some size reduction methods are not economically feasible because of high-energy requirement. Thus, extrusion is a novel and promising method for biomass conversion to biofuel production.

Type of pretreatment	Main characteristics
	In extrusion, materials are subjected to heat, mix and shear, which result in physical and chemical modifications in the biomass after passing through the extruder. This method cannot remove lignin and hemicellulose, and it is and energy demand method.
Microwave	Improve cellulose accessibility and reactivity. The investment cost of equipment is high. Microwave pretreatment on plant fiber must be done at high temperatures (>160°C). On the other hand, it is possible to get high yields, for example, 58 g/100 g (sugar mass/switchgrass mass) that is equivalent to 99% of the total sugar amount (Moretti et al., 2014). This method might be combined with some chemical methods such as alkali and/or acid that could improve pretreatment efficiency.
Ultrasonic	Open the crystalline regions of cellulose, decompose lignin molecules, and improve accessibility and chemical reactivity of cellulose. Nevertheless, this method could have a negative effect in the enzymatic hydrolysis.
High-temperature pyrolysis	This method decomposes cellulose rapidly. Two types can be identified, pyrolysis and liquid hot water decomposition. Pyrolysis: cellulose can be decomposed when heated to above 300°C, resulting in the release of gaseous products and production of coke. Liquid hot water decomposition relies on splitting decomposition of O-acetyl and uronic acid substitution on cellulose, which produces acetic acid and other organic acids. It has the disadvantage that has high energy consumption and low productivity.
High energy electron radiation	It is performed emitting rays at the raw material. This method reduces the level of polymerization in cellulose, loose cellulose structure and increase conversion rates of raw materials, promoting enzymatic hydrolysis. Unfortunately, the operating cost is still high.

CHEMICAL PRETREATMENT

The chemical pretreatment is one of the most used method to enhance the removal of lignin and/or hemicellulose to improve the sugar release from cellulose, reducing crystallinity and degree of polymerization (Behera et al., 2014). Chemical methods have the characteristic that could only improve one or some properties in the lignocellulosic material, since each has a special implementation and purpose. Some chemical pretreatment methods have been identified as toxic compound producers for microorganisms, thereby, showing a little disadvantage in their implementation. The most common methods of chemical pretreatment are based on acids, alkali, oxidation, organosolv, ionic liquids, and deep eutectic solvents. Table 5 shows the main characteristics of the chemical pretreatments.

Table 5. Chemical pretreatment and main characteristics (Behera et al., 2014; Chen et al., 2017; Kim et al., 2016; Rabemanolontsoa and Saka, 2016; Rodriguez-Gomez et al., 2012; Satlewal et al., 2018; Sun et al., 2016)

Type of pretreatment	Main characteristics
Acid	Diverse inorganic acids are employed as catalyst such as, sulfuric, phosphoric, hydrochloric, nitric acids, etc. The use of concentrated acid produces high sugars yields, but also some toxic compounds as furfural and 5-hydrometyl furfural (5-HMF), which are derived from pentoses and hexoses sugars; in addition, those acids might be highly corrosive and expensive. Recently some organic acids have also used such as maleic, succinic, oxalic, fumaric, acetic acids, have also been implemented. They have the advantage of being lower energy demand and corrosive, but the reaction rate is slower than inorganic acids.
Alkali	The alkaline methods aim to remove the lignin, increase pore size, surface area, without degrading carbohydrates. This method employs alkali compounds such as, sodium hydroxide, calcium hydroxide, potassium hydroxide, aqueous ammonia, ammonium hydroxide and hydrogen peroxidase. Alkali methods increase the accessibility of hemicellulases and cellulases, due to some salvation and saponification reactions, which cause swelling reducing the polymerization and crystallinity, but those have less sugar degradation.
Oxidation	This method is applied to degrade lignocellulose and/or lignin by an oxidant agent. Some examples of these methods are ozonolysis, wet oxidation and photocatalysis. Ozonolysis does not produce inhibitors, but it is still costly. Wet oxidation uses oxygen in the presence of water at high temperatures and pressure, thus, the operating cost might be also high. Photocatalysis is another pretreatment method that cannot affect the distribution of the final products, but can reduce the reaction time considerably.
Organosolv	Use organic solvents that degrade lignin based on osmosis principle to break and decompose the internal chemical bonds of cellulose and hemicellulose. This method can obtain pure lignin, cellulose, and hemicellulose with almost no structure exchange. Organosolv pretreatment is still considered with high cost with some effects in the environment and fermentation. Some common solvents employed in the method are methanol, ethanol, acetone, ethylene glycol, and furfuryl alcohol.

Type of pretreatment	Main characteristics
Ionic liquids (IL)	Known as "green solvents," IL are a cellulose solvent and consists on large organic cation and small inorganic anions in liquid phase at ambient conditions. Examples of cations combined with inorganic anions include 1-ethyl-3-methylimidazolium chloride, 1-butyl-3-methylimidazolium chloride and 1-allyl-3-methylimidazolium chloride. An example of cation with organic anion is 1-ethyl-3-methylimidazolium acetate. It was found that ILs containing phosphate or chloride groups enhanced the yield of reducing sugar significantly. IL method is still considered with high cost.
Deep eutectic solvents (DES)	A relatively new technology applied to lignocellulosic biomass pretreatment. They are prepared by combining hydrogen bonding donors (such as quaternary ammonium salts, choline chloride) and hydrogen bonding acceptors such as amides, carboxylic acids, and alcohols at moderate temperatures (60°C to 80°C) to form eutectic mixtures. DES are preferred over conventional IL because they are easy to synthesize, stable, cost-competitive, and typically most of them are environmental-friendly. The implementation in lignocellulosic biomass has shown that it is possible to remove lignin and/or hemicellulose as well as some crystallinity reduction. DES has been effective for diverse type of lignocellulosic materials such as, wheat straw, corn cob, switchgrass, etc. Satlewal et al. (2018) made an extraordinary compilation of the state of the art about DES.

Behena et al. (2014) summarize the use of some chemical pretreatments used in lignocellulose.

PHYSICOCHEMICAL PRETREATMENT

The physicochemical pretreatment combines chemical and physical principles. This method is mainly used to dissolve hemicellulose and alter the lignin structure, allowing to the enzymes to degrade the cellulose. The methods included in this classification are steam explosion, liquid hot water, SO_2 explosion, AFEX and CO_2 explosion. The main characteristics of such methods are shown in Table 6.

Table 6. Physicochemical pretreatment and main characteristics (Behera et al., 2014; Chen et al., 2017; Chiaramonti et al., 2017)

Type of pretreatment	Main characteristics
Steam explosion (SE)	It is one of the most used and cost-effective methods for pretreatment of lignocellulosic biomass. SE technology has been tested in lab and pilot plants scales. It is used for hemicellulose solubilization and lignin transformation. This process has lower environment impact, lower requirement for reaction conditions and cost investment, fewer hazards of chemical reagent, and complete sugar recovery compared with other pretreatment methods. Some phenolic compounds are generated when the lignin is broken down. Steam explosion is typically performed at a temperature range of 160–260 °C (corresponding pressure, 0.69–4.83 MPa) for few seconds to few minutes and further the material is exposed to the atmospheric pressure.
Liquid hot water (LHW)	LHW pretreatment is similar to SE except for the use of water in the liquid state at elevated temperatures (160 – 240 °C) instead of steam. LHW results in hydrolysis of lignocelluloses, removal of lignin, rendering cellulose in the biomass more accessible while avoiding the formation of inhibitory compounds that occur at higher temperatures. This pretreatment offers several advantages, as (i) it does not include any catalyst or chemical, (ii) it requires low temperature, (iii) it minimizes degradation products, (iv) it eliminates the requirement of washing step or neutralization, (v) it has low cost of the solvent for large scale applications.
CO_2 explosion	CO_2 explosion pretreatment method is the addition of CO_2 to the steam explosion pretreatment process to form carbonic acid, which can significantly improve the hydrolysis efficiency of hemicellulose. CO_2 explosion has as advantage not producing inhibitors for subsequent hydrolysis and fermentation. CO_2 is a supercritical fluid, thus, this method can remove lignin effectively, and delignification can increase when add to co-solvents.
SO_2 explosion	The SO_2 explosion is an approach similar to the CO_2 explosion. In SO_2 explosion an external acid addition catalyzes the solubilization of hemicellulose, requires low optimal pretreatment temperature, and provides a partial cellulose hydrolysis. In general, SO_2-catalyzed steam explosion is regarded as one of the most effective pretreatment methods for softwood material. Nevertheless, the drawbacks of the process include the high requirements of the equipment and formation of a large amount of degradation compounds when using acids.
Ammonia fiber/freeze explosion (AFEX)	AFEX is a combination of steam explosion and alkali treatment methods, to pretreat raw material in liquid anhydrous ammonia with high temperature (90–100 °C) and high pressure (1–5.2 MPa).

Type of pretreatment	Main characteristics
	The operational parameters in the AFEX process are ammonia loading, temperature, water loading, blow down pressure, time, and number of treatments. The pressure is immediately released, and ammonia is subsequently evaporated, which leads to rapid temperature change, damaged structure, and increased surface area of cellulose, and improved accessibility of enzyme. The main advantage of AFEX is the absence of inhibition substances for microbial fermentation, the hydrolysate can be directly used without further disposal, and the residue of ammonium salt can be used as microbial nutrition. Nevertheless, ammonia should be recycled because of its high cost and volatility, which can reduce costs and avoid damage to the environment.

BIOLOGICAL PRETREATMENT

Biological pretreatments uses microorganisms like bacteria, and fungi. In general, microorganisms alter or degrade lignocellulose extracellularly by secreting hydrolytic enzyme (such as hydrolases); and ligninolytic enzyme, which depolymerizes lignin (Pérez et al., 2002). Enzymes degrades lignin, cellulose, and hemicellulose structure to release simpler molecules, which are used by the microorganisms in the fermentation step. Biological pretreatment has the advantage that they do not produce toxic compounds during degradation. However, biological method has been less investigated due to low industrial significance and limited technological progress (Sharma et al., 2019). Biological pretreatment generally is implemented as a second pretreatment after the one of the previous methods mentioned above. Table 7 shows the source of some enzymes used for lignocelluloic biomass.

Sharma et al. (2019) provides a summary of different types of microorganisms employed for cellulose, hemicellulose and lignin degradation.

Table 7. Biological pretreatment and main characteristics (Pérez et al., 2002; Sharma et al., 2019; Shindhu et al., 2016)

Type of pretreatment	Main characteristics
Bacteria	There are many bacteria employed in the degradation of biomass. Bacteria are most employed for cellulose and hemicellulose conversion. *Cellulomonas fimi and Thermomonospora fusca*, have been mostly employed to produce cellulases. Other examples of bacteria that release enzymes are: *Cellulosilyticum lentocellum, Ruminococcus flavefaciens, Butyrivibrio fibrisolvens, Prevotella ruminicola, Clostridium thermocellum, Bacteroides cellulosolvens, Zymomonas mobilis*, etc.
Fungi	Several fungal species can produce extracellular fungal cellulose degrading enzymes, where *Trichoderma reesei* is the most employed enzyme to produce commercial cellulases. Regarding to hemicellulose, xylanases are the most employed enzymes to degrade xylan. Xylanases commercial production employs *Aspergillus niger, Trichoderma reesei, Bacillus* and *Humicola insolens*. There is a classification that includes lignocellulolytic fungi that includes species from the ascomycetes (e.g., A*spergillus sp., Penicillium sp., Trichoderma reesei*), basidiomycetes including white-rot fungi (e.g., *Schizophyllum sp., P.chrysosporium*), brown-rot fungi (e.g., *Fomitopsis palustris*) and few anaerobic species (e.g., *Orpinomyces sp.*).

INHIBITORS PRODUCED IN PRETREATMENT

The inhibitors produced in the pretreatment step can really make a difference in the ABE production, since they can decrease the yield. The process inhibitors that might be produced during the pretreatment and neutralization processes are (Baral and Shah, 2014):

- Weak acids formed during hemicellulose degradation: acetic, levulinic and formic.
- Furan derivatives produced from pentose and hexoses: furfural and Hydroxymethylfurfural (HMF)
- Phenolic compounds formed from lignin degradation: p-coumaric acid, ferulic acid, hydroxybenzoic acid, vanillic acid, syringaldehyde, vanillin, *p*-hydroxybenzaldehyde (Luo et al., 2020),

o- hydroxybenzaldehyde and *m*-hydroxybenzaldehyde (Li et al., 2017).
- Salts formed during acid-base neutralization: sodium acetate, sodium chloride and sodium sulfate.

Table 8. Classification of the type of inhibitors effects in different microorganisms that produce ABE

Type of pretreatment	Raw material	Toxic compounds	Effect	Microorganisms	Reference
Diluted sulfuric acid	Model mixture	Furfural, HMF and acetates	Not inhibit in concentration least than 1.98 g/L of acetates.	*Clostridium beijerinckii* BA011	Ezeji et al. (2007).
Diluted sulfuric acid	Wheat straw hydrolysate	Furfural and hydroxymethyl furfural (HMF)	Both furfural and HMF enhanced specific productivity (233–308%) of ABE	*Clostridium beijerinckii* P260	Qureshi et al. (2012)
Diluted sulfuric acid	De-oiled rice bran	furfural, HMF, acetic acid, formic acid and levulinic acid	1 g/L formic acid resulted in total growth inhibition. Detoxification was don using Activated charcoal.	*Clostridium acetobutylicum* YM1	Al-Shorgani et al. (2019)
Sulfuric acid	Corn fiber	--	Do not show any cell inhibition since fermentation inhibitors were removed through a nonionic polymeric adsorbent resin.	*Clostridium beijerinckii* BA101	Qureshi et al. (2008)
Sodium hydroxide	Corn stover / cane molasses	--	For removing the inhibitors pretreated product was autoclaved at 120 °C for 2 h-	*Clostridium saccharobutylicum* DSM 13864	Ni et al. (2013)

Table 8. (Continued)

Type of pretreatment	Raw material	Toxic compounds	Effect	Microorganisms	Reference
Wet disk milling	Corncob	None	Sugar yields of 71.3% for glucose and 39.1% for xylose were observed after enzymatic hydrolysis.	*Clostridium acetobutylicum* SE-1	Zhang et al. (2013)
steam-exploded	Corn stover	Soluble lignin compounds (SLC)	Inhibition when SLC higher than 1.77 g/L	*Clostridium acetobutylicum* ATCC 824	Wang and Chen (2011)
-	Model mixture	Formic acid	At low concentrations (4 g/L) generates the rapid production of acetic and butyric acids and stopping ABE production.	*Clostridium acetobutylicum* DSM 1731	Wang et al. (2011b)
SO_2–ethanol–water (SEW)	spruce chips	--	No apparent inhibitory effect	*Clostridium acetobutylicum* DSM 792	Survase et al. (2011)
Deep eutectic solvent (foraceline with sodium carbonate)	Rice straw	-	No apparent inhibitory effect	*Clostridium saccharobutylicum* DSM 13864	Xing et al. (2018)
Cooked and use of alkaline process / Sulfuric acid	Wood pulp mill	ferulic acid and p-coumaric acid (<0.001 g/L) and significant amounts of furfural, HMF, acetic acid, formic acid, phenolics and levulinic acid.	Presence of inhibitors might reduce 104% buntal production.	*Clostridium beijerinckii* CC101	Lu et al. (2013)

The ABE production is commonly performed by bacteria from *Clostridium* genus, which can be inhibit by several compounds generated due to the type of pretreatment and the raw material. Table 8 classifies the type of inhibitor, concentrations, and species of microorganisms that produce ABE. A more detailed description of the ABE fermentation will be presented in the following section.

The side products obtained in the pretreatment, has opened the opportunity to look for its separation and purification, since some of them have competitive prices in the market. For example, the purification of levulinic acid using intensified processes (Alcocer-García et al., 2019; Alcocer-García et al., 2020) and the purification of furfural (Contreras-Zarazúa et al., 2019; Romero-García et al., 2019).

ABE FERMENTATION

In traditional acetobutyl fermentation, ethanol and acetone are produced in addition to butanol. For this reason, this procedure is known as ABE or acetobutyl fermentation. Butanol production through biomass can be carried out by fermentation with different strains, *Clostridium* microorganisms, e.g., *C. beijerinckii, C. acetobutylicum, C. butylicum* and *C. saccharobutylicum* (Zverlov et al., 2006). Strains of the genus *Clostridium* have the characteristic that they are gram-positive, strictly anaerobic, heterofermentative, and mesophilic. These bacteria have an elongated shape during the exponential growth phase and oval shape when adverse conditions occur. The different clostridial strains that produce butanol are capable of consuming sucrose, fructose, glucose, mannose, lactose, dextrin, starch, glycerol, pentoses (xylose, arabinose) and inulin (Jones and Woods, 1986). Table 9 shows the most relevant strains found in literature, and the preference of the different cultures for ABE fermentation over different hexoses and pentoses (Jones and Woods, 1986, Ezeji et al. 2008).

Table 9. Sugars preference during fermentation in batch reactors for different *Clostridium* cultures (Ezeji et al., 2007, 2008)

Organism	ABE [g·L^{-1}]	Productivity [g·L^{-1}·h^{-1}]	Sugars preferences	% used of glucose: mannose: aribose: xylose
C. acetobutylicum 260	20.3	0.34	Glucose> mannose> arabinose	100:100:92:71
C. butylicum NRRL 592	19.7	0.33	Glucose> mannose> arabinose> xylose	74:38:63:42
C. acetobutylicum 824	18.4	0.19	Glucose> mannose> xylose> arabinose	100:100:60:80
C. beijerinckii BA101	18.0	0.25	Glucose> xylose> arabinose> mannose	100:59:65:78
C. beijerinckii 8052	14.6	0.24	Glucose> xylose> arabinose> mannose	81:42:49:64
C. saccharobutylicum 262	14.3	0.20	Glucose> arabinose> xylose> mannose	74:59:65:78

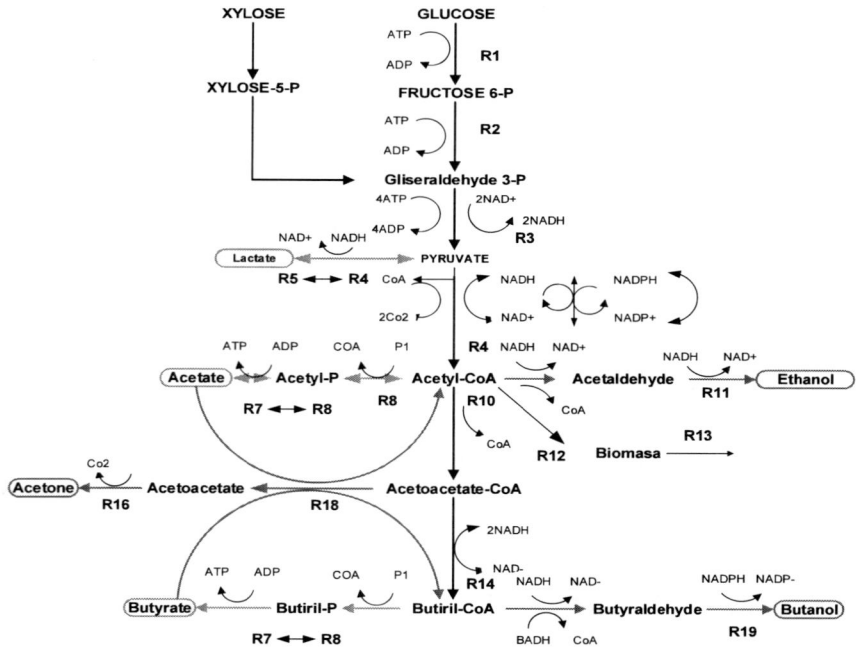

Figure 3. Metabolic pathway of *C. acetobutylicum* using glucose and xylose as substrate (Jones and Woods, 1986).

The *C. acetobutylicum* 260 strain shows the highest productivity of ABE solvents, as well as the highest consumption of sugars. The highest consumption of xylose is presented by the microorganism *C. butylicum* NRRL 592, which refers to an advantage when consuming xylose and glucose. Similarly, it is important to highlight the microorganisms that tolerate and have higher productivity, such as *C. beijerinckii* BA101 (Ezeji et al., 2007). These properties are the reason for the preference over a set of given microorganisms because, with them, a good performance is guaranteed. The metabolic pathway of *C. acetobutylicum* can be seen in Figure 3.

The most recent investigations show that it is possible to produce butanol, by inserting genes from ABE solvent-producing *Clostridium* microorganisms in non-producing microorganism strains, as *Lactobacillus brevis* (Winkler et al., 2011; Berezina et al., 2010), *Pseudomonas putida* (Rühl et al., 2009; Nielsen et al., 2009), *S. cerevisiae* (Steen et al., 2008). *Scherichia coli* (Shen et al., 2011; Baez et al., 2011; Reyes et al., 2011; Inui et al., 2008; Winkler et al., 2010; Atsumi et al., 2008), and *Bacillus subtilis* (Nielsen et al., 2009). Strains of the genus clostridia produce concentrations of butanol less than 2%, while modified non-traditional strains can produce concentrations of up to 3% (Knoshaug and Zhang, 2009; Shen et al., 2011). These strains mainly produce n-butanol or iso-butanol in combination with acetone and butanol. The main disadvantage of the genetically modified butanol-producing strains is their lower ability to metabolize the variety of sugars consumed by *Clostridium* (Green, 2011; Donaldson et al., 2016).

The microorganism *C. acetobutylicum* involves more than 90 genes in the breakdown of carbohydrates (Nölling et al., 2001) and has two main metabolic phases. In the first phase, called acidogenic, the normal, exponential growth of the bacterium occurs, intermediates are produced, mainly organic acids (acetic, butyric and lactic acids), ATP (3.25 mol/mol glucose) and hydrogen. In a second phase, called solventogenic, the stationary decrease stage occurs, the acids generated are consumed, and the main solvents are produced: butanol, acetone, and ethanol, in addition to the synthesis of ATP it is reduced to 2 mol/mol of glucose. The pH during acidogenic can be changed and drop from an initial value of 6.8-7.0 to close to 5.0-4.5 and can be increased in solvent genesis to a pH of 7.0. This metabolic change is explained as an adaptive response of the cell to the low pH of the medium, generated by the presence of acids (Gottwald and Gottschalk, 1985).

An important factor to consider is the accumulation of the main acids and solvents in the fermentation process since they are toxic to the microorganism and eventually results in total inhibition of the metabolism. The change in metabolism from acidogenic to solventogenic is practical to mitigate the toxicity of the main acids. However, the butanol generated inhibits growth at 12 to 16 g·L^{-1} (Woods, 1995). In *C. acetobutylicum*, the toxic effects of butanol are not apparent at concentrations below 4 to 4.85 g·L^{-1} (Jones and Woods, 1986). The presence of ethanol and acetone reduces the growth of the microorganism by approximately 50% at a concentration of around 40 g·L^{-1}, and total inhibition of growth occurs at a concentration of close to 70 g·L^{-1} of acetone and 50 to 60 g·L^{-1}. of ethanol (Bowles and Ellefson, 1985). Among the produced solvents, butanol is the most toxic, being the one reaching inhibitory concentrations during fermentation. On the other hand, raising the concentration of butanol is a key aspect of the energy performance of the process. Studies on the separation of butanol by distillation indicate that if there is an increase in the concentration of butanol in the fermenter from 10 to 40 g·L^{-1}, the use of fuel for the separation is reduced by 80% (Reyes et al., 2011).

Clostridium can produce butanol from different raw materials, including molasses, permeate whey, and corn (Jones and Woods, 1986; Maddox et al. 1993; Ezeji et al. 2007, 2008, Lee et al., 2008, Qureshi et al., 2008). Corn starch and sucrose from molasses are hydrolyzed by clostridia cultures, thereby eliminating the essential hydrolysis step in ethanol production by fermentation. The typical content of lactose in the permeated serum is 44-50 g·L^{-1}. It is suitable for the production of butanol by fermentation, since the toxicity of the product limits feed concentrations to less than 65 g·L^{-1} (Qureshi et al. 2005), being an important factor to take into account for future developments. The whey is rich in minerals, and therefore no mineral supplements are required. The presence of vitamins and micronutrients in molasses makes the fermentation of this substrate preferable to that of corn.

Table 10 shows the performance of butanol production by fermentation for different types of strains using molasses as a substrate. The strain that allows a better yield and concentration of solvents is *C. acetobutylicum* PCISR-10.

Table 10. Fermentation of cane molasses in discontinuous reactors with different strains (Syed, 1994; Shaheen et al. 2000)

Strain	ABE [g·L^{-1}]	Productivity [g·L^{-1}·h^{-1}]	ABE ratio	Yield (g/g)
C.beijerinckii NCP P260	21.9	-	-	33.4
C. saccharobutylicum BAS/B3/SW/336(S)	19.6	-	-	30
C.acetobutylicum PCISR-10	19.2	0.42	1.8:95.3:2.9	34
C. saccharobutylicum NCP P108	18.6	-	-	28.6
C. saccharobutylicum NCP P258	18.3	-	-	30.5
C. saccharoperbutylacetonicum N1-504	15.6	-	-	26
C. acetobutylicum PCISR-5	15.2	0.24	5.3:79:15.7	30
C. acetobutylicum ATCC 4259	9.5	-	-	15.8
C. acetobutylicum ATCC 824	7.8	-	-	13

Different mathematical models have been used to adequately describe the metabolism of fermentation (Lee et al. 2008; Novozymes, 2008; Green et al. 2011, 2016; Donaldson et al., 2016). Shinto et al. (2007, 2008) proposed an on-off metabolic model to describe the consumption of glucose and xylose.

The cost of the substrate is the most important factor in the economy in ABE fermentation (Jones and Woods, 1986; Qureshi et al., 2001). For example, if the yield of butanol can be improved by 19% (from 0.4 to 0.5 g of butanol per g of glucose), the cost of producing butanol is reduced by 14.7% (from 0.34 US to 0.29 $/kg) (Qureshi et al. 2001). For this reason, a wide variety of alternative carbohydrates have been proposed for the production of butanol, such as agricultural residues including corn stubble, corn fiber, and distillery dry grains with corn (Ezeji et al., 2007, 2008, Koukiekolo et al., 2005); artichoke (Marchal et al. 1985); apple pulp (Voget et al. 1985); peanuts (Jesse et al. 2002); wheat straw (Marchal et al. 1986); barley straw (Qureshi et al. 2010); microalgae (Hurun et al. 2010, Nakas, et al. 1983); glycerol (Andrade and Vasconcelos, 2003); potato (Gutierrez et al. 2010); and. organic waste domestic waters (López-Contreras et al. 2000, Claassen et al. 2000).

Table 11 shows the performance of different types of fermentation systems. In a conventional fermentation process, the substrates and nutrients are carried out in batch guards, and they are fed to the reactor at a concentration between 60 and 80 g·L^{-1}.

Table 11. Performance of different ABE fermentation schemes

Process	Productivity ABE (g·L^{-1}·h^{-1})	Yield (g·g^{-1})	ABE concentration (g·L^{-1})	Annotation	Reference
Discontinuous Reactor	0.34	0.42	24.2	59.8 g·L^{-1} glucose fed	Qureshi, et al., 1999 (a)
Discontinuous reactor with pervaporation	0.69	0.42	51.5	78.2 g·L^{-1} glucose fed	Qureshi, et al., 2000 (a)
Pervaporation fed batch reactor	0.98	0.43	165.1	77% glucose conversion	Qureshi and Blaschek, 2000 (a)
Discontinuous reactor with stripping gas	0.61	0.47	75.9	161.7 g·L^{-1} glucose fed	Ezeji et al., 2003 (a)
Discontinuous reactor feed with stripping gas	1.16	0.47	233	500 g·L^{-1} glucose fed	Ezeji et al., 2004 (a)
Continuous reactor with entrained gas	0.91	0.4	46	161.7 g·L^{-1} glucose	Ezeji et al., 2004 (a)
Continuous membrane reactor with liquid liquid extraction	3.08	0.3	14.5	7 g·L^{-1} of biomass	Eckert and Schügerl, 1987 (b)
Integrated fermenter with pertraction	0.21	0.44	9.75	227 g·L^{-1} of lactose fed	Qureshi and Maddox, 2005 (b)
Integrated fermenter with pertraction	0.24	0.37	58	157 g·L^{-1} of used lactose	Tashiro et al., 2005 (b)
Reactor with cell recirculation	6.5	0.35	13	20 g·L^{-1} of biomass	Tashiro et al., 2005 (b)
Reactor with biomass recirculation and cell purge	7.65	-	8.58	200 h of operation	Tashiro et al., 2005 (c)
Continuous two-stage system	0.56	-	21	Dilution rate 0.08 h^{-1}	Godin and Engasser, 1990 (b)
Process	Productivity ABE (g·L^{-1}·h^{-1})	Yield (g·g^{-1})	ABE concentration (g·L^{-1})	Annotation	Reference

Type				Condition	Reference
Continuous membrane reactor with cell recirculation	4.5	-	7	Dilution rate 2 h^{-1}	Afschar et al., 1985 (b)
Continuous reactor immobilized in chitosan	1.43	0.30	2.7	Dilution rate 0.53 h^{-1}	Frick and Schügerl, 1986 (b)
Continuous reactor immobilized in clay	15.8	0.38	7.9	Dilution rate 2 h^{-1}	Qureshi et al., 2000 (a)
Continuous reactor immobilized in trickle bed	4.20	0.34	15.5	Dilution rate 0.27 h^{-1}	Park et al. 1991 (b)
Continuous reactor immobilized in bonechar	6.5	0.38	6.5	Dilution rate 1 h^{-1}	Ramey, 1998 (b)
Continuous reactor immobilized on cotton towel	7.6	0.53	8.5	With butyric acid feed	Huang et al., 2004 (c)
Continuous reactor immobilized on corn stalk	5.06	0.32	5.1	Dilution rate 1 h^{-1}	Zhang et al., 2009 (a)
Continuous reactor with cell recirculation	9.73	0.43	11.5	Concentration glucose, butyrate fed 20, 10 g·L^{-1}, respectively	Baba et al., 2012 (c)

Strains used: a, C. beijerinckii BA101; b, C. acetobutylicum; c, C. saccharoperbutylacetonicum.

The fermentation in batch guards has a duration of 48 to 96 h. The reaction mixture is autoclaved at 121°C for 15 minutes, followed by cooling to 35-37°C and inoculation with the strains of the culture. During cooling, N$_2$ or CO$_2$ is bubbled through the surface to maintain an anaerobic medium. A maximum ABE concentration has been achieved in a batch reactor of 33 g·L^{-1} and a yield of 0.4-0.422 g·g^{-1} with *C. beijerinckii* BA101 (Formanek et al., 1997; Chen and Blaschek, 1999). At the end of fermentation, the cell mass and other suspended solids are removed by centrifugation and sold as livestock feed.

To improve productivity by reaching higher biomass concentrations, techniques such as cell immobilization or recirculation have been proposed (Afschar et al., 1985; Yang and Tsao, 1997; Qureshi et al., 2000; Lienhardt et al., 2002; Tashiro et al., 2005;). Through cell immobilization, maximum ABE productivity of 15.8 g·L^{-1}·h^{-1} has been achieved. However, low substrate conversions were obtained (33%) and an ABE concentration of 7.9 g·L^{-1} (Qureshi et al., 2000). This reactor proved to be stable for about 550 hours. With cell recirculation, by microfiltration and purging, maintaining biomass concentrations around 30 g·L^{-1}, an ABE productivity of 7.65 g·L^{-1}·h^{-1} has been achieved in an operation period of 207 h, with an ABE concentration of 8.58 g·L^{-1} (Tashiro et al., 2005). Although recirculation and immobilization improve productivity between 16 and 30 times compared to a traditional ABE fermenter, other important parameters of the fermentation decrease, such as product concentration and substrate conversion. One way to increase product concentration and substrate conversion is to use fermentation in two or more stages. This fact allows acidogenic and solvatogenic to occur in two separate bioreactors, decreasing cell degeneration. In a two-stage system, a solvent concentration of 18.2 g·L^{-1} and productivity of 1 g·L^{-1}·h^{-1} was reported using *C. acetobutylicum* DSM 1731 (Bahl et al., 1982), that is, the productivity of a discontinuous reactor was doubled, with a product concentration equivalent to the batch process.

PURIFICATION TECHNOLOGIES

Product purification still represents a huge challenge associated with the production of biofuels due to the low concentration obtained from fermentation. In a classical mixture from fermentation an acetone/butanol/ethanol ratio of 3:6:1 is presented; on the other side, the amount of water depends on several factors associated with the upstream process. As example, Jin et al. (2011) reported biobutanol productivity by strain fermentation of 5-10 wt%. As a consequence of those diluted broths, great energy consumption could be required during the purification and separation process. It is desirable a recovery technique with thermal stability, high selectivity and an appropriate removal rate (García et al., 2011). In addition to the difficulty of separating such low biobutanol concentration, the interaction of the components in an ABE mixture turns the separation process more challenging. To exemplify this, observe the ternary diagram in Figure 4, which shows two azeotropes, a homogeneous azeotrope between ethanol and water, and a heterogeneous azeotrope between biobutanol and water.

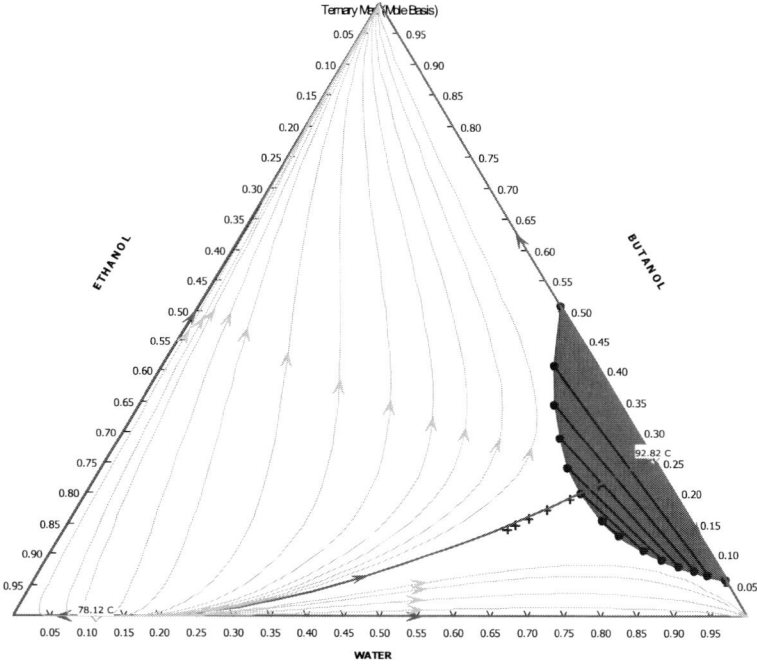

Figure 4. Ternary diagram for the acetone-butanol-ethanol mixture (mole basis).

Nowadays several recovery techniques have been suggested, currently, from those techniques with more investigation, we could highlight distillation, adsorption, liquid-liquid extraction, membrane distillation, and gas stripping.

CONVENTIONAL DISTILLATION AND DISTILLATION-BASED TECHNIQUES

It is not difficult to realize the maturity of distillation and its popularity for mixture separation. With this in mind, several research groups have proposed it as a possible solution to purify the ABE mixture.

Mariano et al. (2011) proposed a flash fermentation integrated with a continuous process for ABE recovery. In their proposal, a bioreactor generates a circulating broth to a vacuum chamber. As a result, this process allows obtaining 30-37g·L^{-1}. Luyben (2012) has proposed a pressure-swing azeotropic distillation system to recover biobutanol from a butanol-water mixture. In this work, two columns with a different range of pressures are used to separate the liquid mixture.

Matsumura et al. (1988) have used distillation columns for biobutanol recovery. This research group reported the energy requirements of 79.5 MJ·kg^{-1} to separate the mixture. Van der Merwe et al. (2013) have reported a set of four alternatives to purifying all the components in the ABE mixture. Most of the schemes are based on distillation columns. As aforementioned, distillation is quite robust for ABE fermentation products recovery, however, using this technique alone is energy-intensive. It has been assumed that producing 1 ton of solvent requires 12 tons of steam (Green, 2011).

Pure distillation is currently the most popular technique for the separation process; however, for ABE fermentation product recovery, distillation possesses several disadvantages such as high energy requirements and low selectivity. With this in mind, enhanced separation techniques have been investigated.

Hybrid processes have been developed to mitigate the energy requirements in the ABE purification step. A relatively well-studied alternative is to combine conventional distillation schemes with liquid-liquid extraction. Liquid-liquid extraction seems an interesting option since an organic water-immiscible extractant agent can be used to avoid the formation of both azeotropes. The organic phase containing biobutanol and the other interest compounds jointly with the extractant agent is further sent to the recovery and purification stage, as shown in Figure 5. The main issue to be solved in liquid-liquid extraction is the correct selection of the extractant agent, which must have high selectivity, and, at the same time, must be harmless for the microorganism involved in the ABE fermentation.

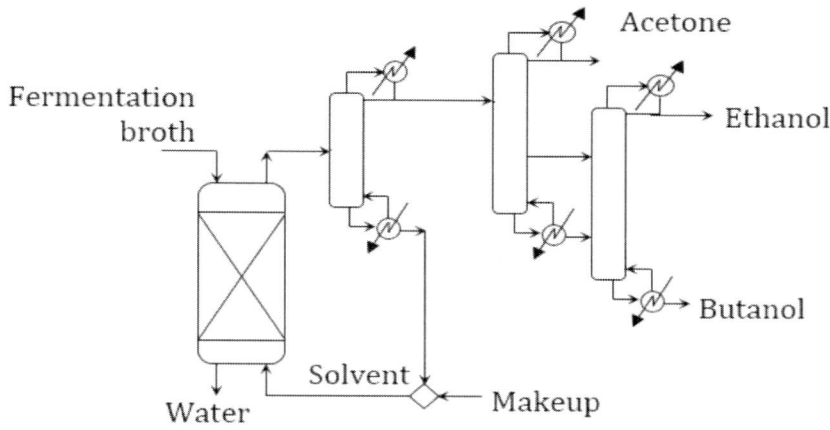

Figure 5. Hybrid liquid-liquid extraction/distillation separation system for the ABE mixture.

MEMBRANE DISTILLATION

Membrane distillation, in general terms, is a separation process in which a hydrophobic membrane is used to separate a fluid mixture with the presence of water (Banat and Al-Shannag, 2000). This type of process is relatively similar to distillation since it requires the feed stream to be heated; this heating evidently will increase latent heat vaporization and consequently the separation will be enhanced. According to the state of the art regarding membrane distillation, there are a variety of membrane distillation modes: osmotic membrane distillation, sweeping gas, air gap, vacuum, and direct contact.

Regarding the ABE mixture, Banat and Al-Shannag (2000) experimentally studied the separation of an ABE mixture using an air gap membrane distillation. According to this research, biobutanol was the component with a higher remotion rate; additionally, the selectivity of ethanol and acetone increase with temperature. Rom et al. (2014) developed a model of vacuum membrane distillation by means of Aspen Plus, this model was developed based on experimental data with poly(propylene) membrane in contact with a biobutanol-water mixture.

Adsorption

Understanding adsorption as a superficial process in which particles from a fluid are attached on a solid surface (Berk, 2009); Lin et al. (2012) reported the use of macroporous resin (KA-I) immersed in polystyrene for the adsorption of the ABE mixture. As a result of their approach, this research group reported an adsorption rate of 139 mg·g^{-1} at 10°C and 304 mg·g^{-1} at 37°C. In other interesting work, by using zeolites in the preparation of mixed membranes, Sharma and Chung (2011) reported an adsorption capacity of 222.24 mg·g^{-1} at 30°C. Similarly, Oudshoorn et al. (2009) reported adsorption of biobutanol over silica zeolites (CBV28014, CBV811C-300, and CBV901), reporting affinity towards biobutanol in the water below 2g·L^{-1}.

On the other hand, Wiehn et al. (2014) adsorbed biobutanol from an ABE broth *in situ* using a hydrophobic resin. Once the process finishes, they produced 27.2 g·L^{-1} and 40.7 g·L^{-1} of solvents and biobutanol, respectively.

Raganti et al. (2020) tested Amberlite XAD-7 as an adsorbent for biobutanol recovery; additionally, they proved the same adsorbent for a wide variety of components such as acetone, biobutanol, ethanol, acetic and butyric acid with an adsorption capacity of 17.1, 102.1, 4.2, 14.1 and 21.3 mg·g^{-1}, respectively.

Adsorption requires few energy investments and has relatively high selectivity, however, it also presents some issues during ABE recovery such as the need for desorption for previously adsorbed components; consequently, additional separation methods must be applied after adsorption.

Liquid-Liquid Extraction

Liquid-liquid extraction (LLE) is a separation process that uses the difference in miscibility to extract a substance from a fluid mixture to a certain extractant agent. Eckert and Schügerl (1987) reported the use of n-decanol to extract biobutanol in a membrane reactor; as a result, they obtained a concentration of biobutanol of 8 g·L^{-1} and productivity of 0.51 g·L^{-1}. However, using n-decanol as extractant produced strain poisoning, reducing the fermentation capacity. In order to avoid poisoning problems, Evans and Wang (1988) used oleyl alcohol as a solvent; in the same sense, Sánchez-Ramírez et al. (2015) reported the use of n-hexyl-acetate as a solvent.

Kurkijärvi et al. (2014) proposed the use of 1-decanol, -heptanol, and 1-octanol in a process of continuous extraction of ABE products. With such an extractive process, Evans and Wang (1988) reported an energy consumption lowered to less than 4MJ·kg^{-1}.

LLE shows a great separation capacity, however, it is a technique that should be handled carefully since it is possible the creation of emulsion and extractant fouling; additionally, if phase separation problems appear, considerable contamination of aqueous streams may happen.

GAS STRIPPING

Gas stripping is a technique which allows removing volatile components coming from ABE fermentation; the basis of this method is to sparge gas into the fermenter in order to condensate volatile components which are subsequently recovered from the condenser.

Park et al. (1991) used a gas-continuos immobilized cell reactor-separator to produce biobutanol from ABE fermentation, they reported a glucose conversion of 30.36 g·L^{-1}. On the other hand, Ezeji et al. (2003) experimentally studied gas-stripping in an ABE batch fermenter, producing an average rate of 17.7 g·L^{-1}. Additionally, they concluded that gas-stripping increases the selectivity of the ABE components and promotes acid assimilation in the fermentation broth. Setlhaku et al. (2013) coupled an ABE fermentor with gas stripping where ABE vapors were stripped using a mixture of 50:50 ethyl glycol-water. The authors reported production of 72.9 g·L^{-1} of acetone, biobutanol, and ethanol.

Table 12. Energy requirements of a variety of purification technologies

Recovery system	Energy Requirement (MJ/kg-product)	Reference
Distillation	12.8	Ruggeri et al. (2015)
Distillation	16.7	Ruggeri et al. (2015)
Distillation	15.2	Ruggeri et al. (2015)
Steam Distillation	21	Qureshi et al. (2005)
Gas stripping	18.9	Qureshi et al. (2005)
Adosrtion	7.1	Qureshi et al. (2005)
Pervaporation	11.9	Qureshi et al. (2005)
Liquid-liquid extraction	7.7	Qureshi et al. (2005)
Vacuum evaporation	21.8	Qureshi et al. (2005)
Double effect Distillation	8	Grisales Díaz and Olivar Tost (2016a)
Pervaporation	9.6	Grisales Díaz and Olivar Tost (2016a)
Vacuum evaporation	10.8	Grisales Díaz and Olivar Tost (2016a)
Adsorption	36.7	Groot et al. (1992)
Gas stripping	23.3	Groot et al. (1992)
Liquid-liquid extraction	15.6	Groot et al. (1992)
Pervaporation	10	Groot et al. (1992)
Distillation	5.2	Grisales Díaz and Olivar Tost (2016a)
Double effect Distillation	3.4	Grisales Díaz and Olivar Tost (2016a)
Vapor compression Distillation	2.5	Grisales Díaz and Olivar Tost (2016a)
Azeotropic Distillation	21.8	Atsumi et al. (2008)
Vapor compression Distillation	3.7	Grisales Díaz and Olivar Tost (2016a)
Double effect Distillation	5.7	Grisales Díaz and Olivar Tost (2016a)

The gas-stripping technique presents several advantages over other methodologies, such as relatively simple operation, low cost, and relatively good efficiency. However, it also presents disadvantages as the formation of foam in the reactor, so there is a need for an antifoam agent; this may also turn into a low ABE productivity.

Table 13. Advantages and disadvantages of technologies for biobutanol purification

Technology	Advantage	Disadvantages	Reference
Distillation and distillation-based	The ability to handle a wide range of feed flow rates	High energy usage	Smith and Jobson (2000)
	The ability to separate feeds with a wide range of feed concentrations		
	The ability to produce high product purity		
Adsorption	Adsorbents used in adsorption technique possess relative high selectivity towards butanol over water	Waste production	Sadegh and Ali (2019)
	Low-cost	Difficulties in desorption of organic compound previously adsorbed on the sorbent	
	Easy Operation		
Liquid-Liquid Extraction	High capacity of the extractantion	Application of good extractants to ABE fermentation broth for direct removal of n-butanol can cause destruction of bacteria strains in fermenter	Wu and Tu (2016)
	High selectivity of separation	Generation of emulsions and extractant fouling	
Membranes	It can be produced with extremely high selectivities	Membrane processes rarely allow obtaining more than a single product with high purity	Saleh and Gupta (2016)
	They are potentially better for the environment since the membrane approach require the use of relatively simple and non-harmful materials	Membrane processes cannot be easily staged	
		Membrane processes often do not scale up very well to accept massive stream sizes	
Gas stripping	It is simple and inexpensive to operate	The tiny bubbles, produced in gas stripping, create excessive amounts of foam in a bioreactor	Kujawska et al. (2015)
	It does not suffer from fouling or clogging due to the presence of biomass	The need of adding an antifoam agent, which can be toxic to bacteria	

Table 12 summarizes some technologies for biobutanol purification, highlighting the energy requirements per kg of ABE products. Some techniques have been already discussed so far, many others are only for comparison purposes. Additionally, Table 13 shows the advantages and disadvantages of the separation technologies used in the production of biobutanol.

CURRENT STATE OF INDUSTRIAL PRODUCTION

From the first production of butanol with bacteria by Louis Pasteur in 1862, the production of butanol, the fermentation as a way to produce butanol or acetone in great scale started gaining interest (Dürre, 2008). Industrial ABE fermentation was widely exploited in the 40's, contributing with around two-thirds of the butanol production in U.S.A. by 1945 (Birgen et al., 2019). Before the end of the 40's, 66% of the butanol used in the world was produced by fermentation (Rose, 1961). Nevertheless, as the petroleum industry grew up, offering butanol and other products at low prices, the use of ABE fermentation to produce that alcohol decayed. Nowadays biotechnological paths have regained interest from an industrial point of view. According to the U.S. Department of Energy (n.d.) and Moriarty et al. (2020), two of the main companies producing biobutanol are Butamax and Gevo. Both companies produce biobutanol from corn in retrofitted bioethanol production facilities. On the other hand, the biobutanol produced bu Gevo is further treated to produce biojet fuel. Ramos et al. (2016) reported that the production of butanol by Gevo by 2014 was 50,000 gallons. Ni and Sun (2009) report that ABE facilities in China re-started production in 2006, having a growing expectation up to 1,000,000 tons of acetone, butanol and ethanol. An example of these chinese facilities is given by Cathay Industrial Bio, located in Jilin, which produce 100,000 tonnes per year of biobutanol from corn (Loneylinked, 2017). Green (2011) and Mariano et al. (2013) reported the opening of a biobutanol plant in Brazil, producing 8,000 ton of ABE mixture from sugar cane juice. According to Bankar et al. (2013), most of the ABE fermentation plants work in semicontinuous operation and are located close to biethanol facilities to reduce operating costs.

It is clear that industrial production of biobutanol has regained interest in the last two decades, nevertheless, its production is still not competitive with the butanol with fossil origin. Because of this, the industrial production of butanol by ABE fermentation has been limited to a few countries, mainly those with an abundance of sugar/starchy raw materials. There are many areas of opportunity which need to be taken care of to promote the production of this biofuel. This topic will be covered in the next section.

AREAS OF OPPORTUNITY

Although the conversion of biomass into is technically feasible, there are still many areas of opportunity to allow its production in industrial scale to be economically feasible, turning the biobutanol into a competitive fuel. Firstly, the explotation of all the potential biobutanol sources is key to reduce the cost associated with raw materials. As for many biofuels, the cost of raw materials represents a high contribution to the production cost of biobutanol. The data for industrial production of butanol indicates that it is obtained mainly from first-generation materials. Thus, the use of second-generation materials, as agricultural residues, must be promoted, since it gives an additional use to the residues, with the expectation of paying lower quantities for such raw materials. Of course, the use of such residues implies that enhancements must be developed for the pretreatment and hydrolysis steps, to obtain higher yields. Rotten fruits can also be an important source of sugars, together with the residues from the confectionery industry. In the case of lignocellulosic materials, market residues and industrial residues, one of the main challenges relies on the development of a proper supply chain, since the cost of collecting those residues from the produces could be high if the collecting strategy is inappropriate. Another kind of raw material that is interesting for the production of butanol is microalgal biomass since they have a high content of carbohydrates and require relatively low time to grow. Nevertheless, the high capital and operating costs in microalgae cultivation are the main issues to be solved to turn such biomass into an economically feasible alternative (Wang et al., 2017).

Another hurdle in the biobutanol production is the low concentration and yield, together with the inhibition of the microorganisms due to butanol. Thus, enhancements are required in the fermentation stage, for which genetic engineering has a key role to enhance the butanol yield. Bramucci and Nagarajan (2016) reported the development of recombinant yeast host cells resistant to butanol. Similarly, Lee and Eom (2018) developed butanol-resistant recombinant yeast host cells, mentioning that the microorganisms showed acetyl-CoA and butytyl-CoA synthesis pathways. Nguyen et al. (2018) described a modified *Clostridium acetobutylicum* bacillus to allow the production of n-butanol from glucose in a continuous regime. Such operation mode was difficult to handle due to degeneration issues with the bacillus. Additional to the development of modified microorganisms, alternative production schemes have been proposed to enhance yield and productivity. Besides the genetic modification of strains, Goyal and Khanna (2019) mention that other indigenous microbes, as *Pseudomonas VLB120*, *Enterococcus faecenium*, *Lactobacillus plantarum*, among others, have a high natural resistance to butanol. Nevertheless, the yield to butanol is lower than that reported for clostridial bacillus. For the processing of lignocellulosic biomass, simultaneous saccharification and fermentation have been reported (Ibrahim et al., 2015). Nevertheless, such an approach required a match between the saccharification and fermentation conditions, which may not necessarily occur. Ibrahim et al. (2018) report the consolidated bioprocessing as a technology able to enhance the biobutanol yield; in such approach the enzyme production, saccharification and fermentation take place in the same equipment. This approach seems promissory, but it is necessary to have compatibility between the co-cultures, and there may exist a conflict on the operating conditions.

There are also areas of opportunity in the separation sequence. As aforementioned, the separation of the components in the ABE mixture from water is not an easy task, due to the existence of homogeneous and heterogeneous azeotropes. Additionally, there are high energy requirements in the downstream processes, due to the highly diluted products obtained in fermentation. Thus, the enhancement of the separation train is mandatory, to allow reducing the energy required to perform the separation, but avoiding the use of extremely expensive technologies. In the case of the use of liquid-liquid extraction to separate the water, followed by distillation, as reported by Sánchez-Ramírez et al. (2015) and Errico et al. (2016), the higher the quantity of solvent required for the remotion of water, the higher the energy to be used to recover the solvent in the distillation column. According to the results reported by Contreras-Vargas et al. (2019), the energy requirement of the solvent recovery column may represent more than 95% of the total energy requirement for the purification of the ABE mixture is such extraction-distillation systems. Thus, more selective solvents must be used, to allow removing the water with the lowest quantity of solvent. If less solvent is required, the energy requirements in the recovery column will be reduced, substantially reducing the total energy requirements on the train. To select the solvent, the computer-aided molecular design is a useful tool to explore a wide diversity of solvents. Sánchez-Ramírez et al. (2018) proposed a framework to simultaneously analyze the solvent alternatives and obtaining the optimal design of the purification step. Such tools can be exploited to significantly reduce the use of utilities in the separation of the ABE mixture.

Another area of opportunity is related to the use of ethanol-butanol or ethanol-butanol-gasoline mixtures to run internal combustion engines, instead of butanol-gasoline mixtures, which implies separating the pair ethanol/butanol in the purification step of the ABE fermentation process. Li et al. (2016) studied the effect of using isopropanol/n-butanol/ethanol mixtures diluted in gasoline to run an engine, finding that a 30 vol% for alcohols in gasoline allowed a good performance for the analyzed system. Similarly, Arce-Alejandro et al. (2018) reported the use of ethanol/butanol mixtures in an internal combustion engine, finding that the presence of small quantities of ethanol (up to 20 vol%) had little effect on the engine performance, in comparison with the performance using only n-butanol. From these results, Contreras-Vargas et al. (2019) proposed eliminating the distillation system which separated the ethanol/butanol pair in the ABE fermentation process. Nevertheless, this simplification in the separation train caused only a small reduction in the total annual cost and the environmental impact, since most of the contribution to those indicators is given by the recovery of the solvent, as previously mentioned. Even so, avoiding such separation allows reducing the need for space for a column and the heating/cooling equipment associated. Unfortunately, there are standards for the use of ethanol/gasoline mixtures (ASTM D4806-19a, ASTM D5798-19b) and butanol/gasoline mixtures (ASTM D7862-19), but not for ethanol/butanol or ethanol/butanol/gasoline mixtures. Thus, the development of such standards is an area of opportunity.

Another area of opportunity is the application of process intensification to the different sections of the production process. In general terms, process intensification implies strategies to reduce the size of process equipment, the capital cost and the generation of wastes, improving energy efficiency and inherent safety (Richardson et al., 2002; Qian et al., 2018), contributing to the development of more sustainable processes (Cremaschi, 2014). From these concepts, intensification alternatives have been proposed for the biobutanol production process. For the fermentation section, Masngut and Harvey (2012) tested a batch oscillatory baffled reactor, reporting that configuration allowed increasing the productivity by 38% in comparison with a stirred tank reactor. Eom et al. (2013) performed preliminary studies for an extractive fermentation system. Liu et al. (2014) proposed the incorporation of a pervaporation membrane to the fermentation equipment, reporting that such modification duplicates the ABE productivity. In the case of the purification section, there are different alternatives for intensification which can be explored. Errico et al. (2016) proposed some hybrid separation processes to purify an ABE mixture at high biobutanol concentration; the hybrid configuration is composed of a liquid-liquid extractive column, followed by thermally coupled distillation columns, generating several alternatives with relatively low energy consumption. Sánchez-Ramírez et al. (2017) reported the use of a liquid-liquid extraction system combined with different dividing wall column configurations to separate the ABE mixture, comparing them in terms of total annual cost, environmental impact and controllability. Errico et al. (2017) also proposed a set of hybrid alternatives based on liquid-liquid extractive columns combined with dividing wall columns. Patraşcu et al. (2018) reported the use of an azeotropic dividing wall distillation column with vapor recompression, obtaining a reduction of 58% in energy requirements, in comparison with a conventional distillation/decantation sequence. Contreras-Vargas et al. (2019) reported the use of intensified distillation systems in a liquid-liquid extraction plus distillation scheme to obtain the mixture ethanol/butanol, mentioning that the use of intensified sequences reduces significantly the risk associated with the separation train. Segovia-Hernández et al. (2020) proposed a set of intensified distillation systems following a liquid-liquid extraction column. As part of the results, the authors mention that the intensified structures may help to improve the sustainability and safety of the process, but some configurations may show bad dynamic performance. As can be seen from the previously discussed examples, the development of intensified technologies and its application to the ABE fermentation process can considerably enhance its economical and environmental indicators, increasing the safety of the process due to the reduction in the number of equipment. Although studies on the dynamic properties of the intensified separation systems have been reported, the detailed design of the control structure is still required. Additionally, most of the studies on intensified separation technologies are developed in simulation environments, thus an area of opportunity is the study of the operability of the physical systems. In the case of the experimental proposals, further studies are necessary to test its economic feasibility in an industrial scale.

CONCLUSION

Biobutanol is a biofuel with several advantages over bioethanol to be used as a fuel in gasoline engines, even biobutanol presents a higher octane number than gasoline. Additionally, biobutanol can be used in any blend proportion with gasoline or it can be used pure biobutanol in spark-ignition engines without engine modification. It is produced from a variety of renewable materials, which turns it into an interesting alternative to gasoline, due to the need for cleaner fuels in the transportation sector. A great advantage over other bioalcohols such as ethanol is that biobutanol fermentation is able of using two kinds of sugar, pentoses, and hexoses. The industrial production of butanol through biological pathways has regained interest in the last years, with a particular focus on the ABE fermentation process. Currently, there are still issues to be solved in the biobutanol production, as promoting the use of low-cost wastes as raw materials, developing butanol-resistant microorganisms but keeping or increasing the yield, enhance the separation train to reduce its cost and energy requirements, and promoting the use of mixtures of alcohols to make use of the pair ethanol/butanol as fuel or additives. However, the two key issues in biobutanol production are the connection between fermentation performance and separation process; those two single stages are called to promote a profitable bioutanol production. If bacterias enhance its biobutanol production, the separation units will decrease the energy consumption; consequently, the economics and environmental impact of the entire process will decrease. These developments will be indispensable to spread the production of biobutanol around the world, making use of locally produced non-edibles raw materials.

REFERENCES

Afschar, A. S., Biebl, H., Schaller, K. and Schügerl, K. (1985). Production of acetone and butanol by Clostridium acetobutylicum in continuous culture with cell recycle. *Applied Microbiology and Biotechnology*, 22(6): 394-398.

Al-Shorgani, N. K. N., Al-Tabib, A. I., Kadier, A., Zanil, M. F., Lee, K. M., Kalil, M.S. (2019). Continuous Butanol Fermentation of Dilute Acid-Pretreated De-oiled Rice Bran by Clostridium acetobutylicum YM1. *Scientific Reports*. 9: 4622.

Alcocer-García, H., Segovia-Hernández, J. G., Prado-Rubio, O. A., Sánchez-Ramírez, E., Quiroz-Ramírez, J. J. (2019). Multi-objective optimization of intensified processes for the purification of levulinic acid involving economic and environmental objectives. *Chemical Engineering and Processing - Process Intensification*. 136:123-137.

Alcocer-García, H., Segovia-Hernández, J. G., Prado-Rubio, O. A., Sánchez-Ramírez, E., Quiroz-Ramírez, J. J. (2020). Multi-objective optimization of intensified processes for the purification of levulinic acid involving economic and environmental objectives. Part II: A comparative study of dynamic properties. *Chemical Engineering and Processing - Process Intensification*. 147: 107745.

Andrade, J. C. and Vasconcelos, I. (2003). Continuous cultures of Clostridium acetobutylicum: culture stability and low-grade glycerol utilisation. *Biotechnology Letters*, 25(2): 121-125.

Arce-Alejandro, R., Villegas-Alcaraz, J. F., Gómez-Castro, F. I., Juárez-Trujillo, L., Sánchez-Ramírez, E., Carrera-Rodríguez, M., and Morales-Rodríguez, R. (2018). Performance of a gasoline engine powered by a mixture of ethanol and n-butanol. *Clean Technologies and Environmental Policy*, 20: 1929-1937.

Atsumi, S., Cann, A. F., Connor, M. R., Shen, C. R., Smith, K. M., Brynildsen, M. P., Chou, K. J., Hanai, T. and Liao, J. C. (2008). Metabolic engineering of Escherichia coli for 1-butanol production. *Metabolic Engineering*, 10(6): 305–311.

Baba, Y., Matsuki, Y., Mori, Y., Suyama, Y., Tada, C., Fukuda, Y., Saito, M., Nakai, Y. (2017). Pretreatment of lignocellulosic biomass by cattle rumen fluid for methane production: Bacterial flora and enzyme activity analysis. *Journal of Bioscience and Bioengineering*. 34: 421-428.

Baba, S. I., Tashiro, Y., Shinto, H. and Sonomoto, K. (2012). Development of high-speed and highly efficient butanol production systems from butyric acid with high density of living cells of Clostridium saccharoperbutylacetonicum. *Journal of Biotechnology*, 157(4): 605-612.

Baez, A., Cho, K. M., and Liao, J. C. (2011). High-flux isobutanol production using engineered Escherichia coli: a bioreactor study with in situ product removal. *Applied Microbiology and Biotechnology*, 90(5): 1681-1690.

Bahl, H., Andersch, W. and Gottschalk, G. (1982). Continuous production of acetone and butanol by Clostridium acetobutylicum in a two-stage phosphate limited chemostat. *European Journal of Applied Microbiology and Biotechnology*, 15(4): 201-205.

Balat, M., Balat, H. and Öz, C. (2008). Progress in bioethanol processing. *Progress in Energy and Combustion Science*, 34(5): 551–573.

Banat, F. A. and Al-Shannag, M. (2000). Recovery of dilute acetone–butanol–ethanol (ABE) solvents from aqueous solutions via membrane distillation. *Bioprocess Engineering*, 23:643–649.

Bankar, S. B., Survase, S. A., Ojamo, H. and Granström, T. (2013). Biobutanol: the outlook of an academist and industrialist. *RSC Advances*, 47(3): 24734-24757.

Baral, N. R., Shah, A. (2014). Microbial inhibitors: formation and effects on acetone-butanol-ethanol fermentation of lignocellulosic biomass. *Applied Microbiology and Biotechnology*. 98: 9151–9172.

Berezina, O. V., Zakharova, N. V., Brandt, A., Yarotsky, S. V., Schwarz, W. H. and Zverlov, V. V. (2010). Reconstructing the clostridial n-butanol metabolic pathway in Lactobacillus brevis. *Applied Microbiology and Biotechnology*, 87(2): 635-646.

Behera, S., Arora, R., Nandhagopal, N., Kumar, S. (2014). Importance of chemical pretreatment for bioconversion of lignocellulosic biomass. *Renewable and Sustainable Energy Reviews*. 36: 91-106.

Berk Z. (2009). Adsorption and ion exchange. In: Berk Z. (Ed.), *Food Process Engineering and Technology*. Academic Press, San Diego: 279–294.

Birgen, C., Dürre, P., Preisig, H. A. and Wentzel, A. (2019). Butanol production from lignocellulosic biomass: revisiting fermentation performance indicators with exploratory data analysis. *Biotechnology for Biofuels*, 12: 167.

Bowles, L. K. and Ellefson, W. L. (1985). Effects of butanol on Clostridium acetobutylicum. *Applied and Environmental Microbiology*, 50(5): 1165-1170.

Bramucci, M. G. and Nagarajan, V. (2016). *Butanol tolerance in microorganisms*. U.S. Patent No. 9,237,330 B2.

Brosowski, A., Thrän, D., Mantau, U., Mahro, B., Erdmann, G., Adler, P., Stinner, W., Reinhold, G., Hering, T. and Blanke, C. (2016). A review of biomass potential and current utilisation – status quo for 93 biogenic wastes and residues in Germany. *Biomass and Bioenergy*, 95: 257-272.

Castro-Montoya, A. J. and Jiménez-Gutiérrez, A. (2013). Lignocellulosic biomass: a raw material for the future. In: Stuart, P. R., El-Halwagi, M. M. (Eds.), *Integrated Biorefineries: design, analysis and optimization*. CRC Press, Boca Raton: 461-468.

Chen, C. K. and Blaschek, H. P. (1999). Acetate enhances solvent production and prevents degeneration in Clostridium beijerinckii BA101. *Applied Microbiology and Biotechnology*, 52(2): 170-173.

Chen, H., Liu, J., Chang, X., Chen, D., Xue, Y., Liu, P., Lin, H., Han, S. (2017). A review on the pretreatment of lignocellulose for high-value chemicals. *Fuel Processing Technology*. 160: 196-206.

Chiaramonti, D., Prussi, M., Ferrero, S., Oriani, L., Ottonello, P., Torre, P., Cherchi, F. (2012). Review of pretreatment processes for lignocellulosic ethanol production, and development of an innovative method. *Biomass and Bioenergy*. 46: 25-35.

Choi, I. S., Wi, S. G., Kim, S. B. and Bae, H. J. (2016). Conversion of coffee residue waste into bioethanol with using popping pretreatment. *Bioresource Technology*, 125: 132-137.

Chuck, C. J. (2016). *Biofuels for aviation: feedstocks, technology and implementation*. Academic Press, San Diego.

Claassen, P. A., Budde, M. A. and López-Contreras, A. M. (2000). Acetone, butanol and ethanol production from domestic organic waste by solventogenic clostridia. *Journal of Molecular Microbiology and Biotechnology*, 2(1): 39-44.

Contreras-Vargas, C. A., Gómez-Castro, F. I., Sánchez-Ramírez, E., Segovia-Hernández, J. G., Morales-Rodriguez, R. and Gamiño-Arroyo, Z. (2019). Alternatives for the purification of the blend butanol/ethanol from an acetone/butanol/ethanol fermentation effluent. *Chemical Engineering and Technology*, 42(5): 1088-1100.

Contreras-Zarazúa, G., Jasso-Villegas, M. E., Sanchez-Ramirez, E., Vazquez-Castillo, J. A., Segovia-Hernandez, J. G., (2019). Design and optimization of azeotropic and extractive distillation to purify furfural considering safety, environmental and economic issues, *Computer Aided Chemical Engineering*. 46: 139-144.

Cremaschi, S. (2014). A perspective on process synthesis: challenges and prospects. *Computer Aided Chemical Engineering*, 34: 35-44.

Dautidis, P., Marvin, W. A., Rangarajan, S. and Torres, A. I. (2013). Engineering Biomass Conversion Processes: A Systems Perspective, *AIChE Journal*, 59: 3-18.

Diario Oficial (2008). *Ley de Promoción y Desarrollo de los Bioenergéticos*. [*Law of Promotion and Development of Bioenergetics*].

Donaldson, G. K., Huang, L. L., Maggio-Hall, L. A., Nagarajan, V., Nakamura, C. E. and Suh, W. (2016). *Fermentive production of four carbon alcohols*. U.S. Patent No. 9,297,028.

Drapcho, C. M., Nhuan, N. P. and Walker, T. H. (2008). *Biofuels Engineering Process Technology*. McGraw Hill, U.S.A.

Dürre, P. (2008). Fermentative butanol production: bulk chemical and biofuel. *Annals of the New York Academy of Sciences*, 1125: 353-362.

Eckert, G., & Schügerl, K. (1987). Continuous acetone-butanol production with direct product removal. *Applied Microbiology and Biotechnology*, 27(3): 221-228.

Eom, M. H., Kim, W., Lee, J., Cho, J. H., Seung, D., Park, S. and Lee, J. H. (2013). Modeling of a biobutanol adsorption process for designing an extractive fermentor. *Industrial and Engineering Chemistry Research*, 52(2): 603-611.

Errico, M., Sanchez-Ramirez, E., Quiroz-Ramìrez, J. J., Segovia-Hernández, J. G., Rong, B. G. (2016). Synthesis and design of new hybrid configurations for biobutanol purification. *Computers & Chemical Engineering*, 84, 482-492.

Errico, M., Sanchez-Ramirez, E., Quiroz-Ramírez, J. J., Rong, B. G., Segovia-Hernandez, J. G. (2017). Multiobjective optimal acetone–butanol–ethanol separation systems using liquid–liquid extraction-assisted divided wall columns. *Industrial & Engineering Chemistry Research*, 56(40), 11575-11583.

Evans, P. J. and Wang, H. Y. (1988). Enhancement of butanol formation by Clostridium acetobutylicum in the presence of decanol-oleyl alcohol mixed extractants. *Applied and Environmental Microbiology*, 54(7):1662–1667.

Ezeji, T. and Blaschek, H. P. (2008). Fermentation of dried distillers' grains and solubles (DDGS) hydrolysates to solvents and value-added products by solventogenic clostridia. *Bioresource Technology*, 99(12): 5232-5242.

Ezeji, T. C., Qureshi, N. and Blaschek, H. P. (2003). Production of acetone, butanol and ethanol by Clostridium beijerinckii BA101 and in situ recovery by gas stripping. *World Journal of Microbiology and Biotechnology*, 19: 595–603.

Ezeji, T. C., Qureshi, N. and Blaschek, H. P. (2004a). Butanol fermentation research: upstream and downstream manipulations. *The Chemical Record*, 4(5): 305-314.

Ezeji, T. C., Qureshi, N. and Blaschek, H. P. (2004b). Acetone butanol ethanol (ABE) production from concentrated substrate: reduction in substrate inhibition by fed-batch technique and product inhibition by gas stripping. *Applied Microbiology and Biotechnology*, 63(6): 653-658.

Ezeji, T., Qureshi, N. and Blaschek, H. P. (2007). Butanol production from agricultural residues: impact of degradation products on Clostridium beijerinckii growth and butanol fermentation. *Biotechnology and Bioengineering*, 97(6): 1460-1469.

Formanek, J., Mackie, R. and Blaschek, H. P. (1997). Enhanced butanol production by Clostridium beijerinckii BA101 grown in semidefined P2 medium containing 6 percent maltodextrin or glucose. *Applied and Environmental Microbiology*, 63(6): 2306-2310.

Frick, C. and Schügerl, K. (1986). Continuous acetone-butanol production with free and immobilized Clostridium acetobutylicum. *Applied Microbiology and Biotechnology*, 25(3): 186-193.

García, V., Päkkilä, J., Ojamo, H., Muurinen, E., Keiski, R. L. (2011) Challenges in biobutanol production: How to improve the efficiency? *Renewable and Sustainable Energy Reviews*, 15(2), 964–980.

García-Aparicio, M. P., Ballesteros, M., Manzanares, P., Ballesteros, I., González, A. and Negro, M. J. (2007). Xylanase contribution to the efficiency of cellulose enzymatic hydrolysis of barley straw. *Applied Biochemistry and Biotechnology*, 136: 353-365.

Gibreel, A., Sandercock, J. R., Lan, J., Goonewardene, L. A., Zijlstra, R. T., Curtis, J. M. and Bressler, D. C. (2009). Fermentation of barley by using *Saccharomyces cerevisae*: Examination of barley as a feedstock for bioethanol production and value-added products. *Applied and Environmental Microbiology*, 75(5): 1363-1372.

Godin, C. and Engasser, J. M. (1990). Two-stage continuous fermentation of Clostridium acetobutylicum: effects of pH and dilution rate. *Applied Microbiology and Biotechnology*, 33(3): 269-273.

Goyal, L. and Khanna, S. (2019). Recent advances in microbial production of butanol as a biofuel. *International Journal of Applied Sciences and Biotechnology*, 7(2): 130-152.

Gottwald, M. and Gottschalk, G. (1985). The internal pH of Clostridium acetobutylicum and its effect on the shift from acid to solvent formation. *Archives of Microbiology*, 143(1): 42-46.

Green, E. M. (2011). Fermentative production of butanol – the industrial perspective. *Current Opinion in Biotechnology*, 22(3): 337-343.

Grisales Díaz, V. H. and Olivar Tost, G. (2016a). Butanol production from lignocellulose by simultaneous fermentation, saccharification, and pervaporation or vacuum evaporation. *Bioresource Technology*, 218: 174–182.

Grisales Díaz, V. H., Olivar Tost, G. (2016b). Ethanol and isobutanol dehydration by heat integrated distillation, *Chemical Engineering and Processing: Process Intensification*, 108: 117–124.

Groot, W. J., van der Lans, R. G. J. M. and Luyben, K. Ch. A. M. (1992). Technologies for butanol recovery integrated with fermentations, *Process Biochemistry*, 27(2): 61–75.

Gutierrez, N. A., Maddox, I. S., Schuster, K. C., Swoboda, H. and Gapes, J. R. (1998). Strain comparison and medium preparation for the acetone-butanol-ethanol (ABE) fermentation process using a substrate of potato. *Bioresource Technology*, 66(3): 263-265.

Harrison, S. T. L., Johnstone-Robertson, M., Rademeyer, S., Murhonyi, L, Ngwenya, C., Horn, C., Rumjeet, S and Smart, M. (2019). *Value recovery from solid confectionery waste*. Technical Report. Centre for Bioprocess Engineering Research, University of Cape Town, South Africa.

Harun, R., Singh, M., Forde, G. M. and Danquah, M. K. (2010). Bioprocess engineering of microalgae to produce a variety of consumer products. *Renewable and Sustainable Energy Reviews*, 14(3): 1037-1047.

Hechinger, M., Voll, A. and Marquardt, W. (2010). Towards an integrated design of biofuels and their production pathways, *Computers & Chemical Engineering,* 34: 1909-1918.

Ho, M. C., Ong, V. Z., Wu, T. Y. (2019). Potential use of alkaline hydrogen peroxide in lignocellulosic biomass pretreatment and valorization – A review. *Renewable and Sustainable Energy Reviews*. 112: 75-86.

Huang, W. C., Ramey, D. E. and Yang, S. T. (2004). Continuous production of butanol by Clostridium acetobutylicum immobilized in a fibrous bed bioreactor. In Finkelstein, M., Davison, B. (Eds.) *Proceedings of the Twenty-Fifth Symposium on Biotechnology for Fuels and Chemicals*. Humana Press, New Jersey: 887-898.

Ibrahim, M. F., Abd-Aziz, S., Yusoff, M. E. M., Phang, L. Y. and Hassan, M. A. (2015). Simultaneous enzymatic saccharification and ABE fermentation using pretreated oil palm empty fruit bunch as substrate to produce butanol and hydrogen as biofuel. *Renewable Energy*, 77: 447-455.

Ibrahim, M. F., Kim, S. W. and Abd-Aziz, S. (2018). Advanced bioprocessing strategies for biobutanol production. *Renewable and Sustainable Energy Reviews*, 91: 1192-1204.

Inui, M., Suda, M., Kimura, S., Yasuda, K., Suzuki, H., Toda, H., Yamamoto, S., Okino, S., Suzuki, N. and Yukawa, H. (2008). Expression of Clostridium acetobutylicum butanol synthetic genes in Escherichia coli. *Applied Microbiology and Biotechnology*, 77(6): 1305-1316.

Jesse, T. W., Ezeji, T. C., Qureshi, N. and Blaschek, H. P. (2002). Production of butanol from starch-based waste packing peanuts and agricultural waste. *Journal of Industrial Microbiology and Biotechnology*, 29(3): 117-123.

Jin, C., Yao, M. F., Liu, H. F., Lee, C. F. F., Ji, J. (2011) Progress in the production and application of n-butanol as a biofuel. *Renewable and Sustainable Energy Reviews*, 15(8):4080–4106.

Jones, D. T. and Woods, D. R. (1986). Acetone-butanol fermentation revisited. *Microbiological Reviews*, 50(4): 484-524.

Khalil, S. R. A., Abdelhafez, A. A. and Amer, E. A. M. (2015). Evaluation of bioethanol production from juice and bagasse of some sweet sorghum varieties. *Annals of Agricultural Sciences*, 60(2): 317-324.

Kim, J. S., Lee, Y. Y., Kim, T. H. (2016). A review on alkaline pretreatment technology for bioconversion of lignocellulosic biomass. *Bioresource Technology*. 199: 42-48.

Knoshaug, E. P. and Zhang, M. (2009). Butanol tolerance in a selection of microorganisms. *Applied Biochemistry and Biotechnology*, 153(1-3): 13-20.

Koukiekolo, R., Cho, H. Y., Kosugi, A., Inui, M., Yukawa, H. and Doi, R. H. (2005). Degradation of corn fiber by Clostridium cellulovorans cellulases and hemicellulases and contribution of scaffolding protein CbpA. *Applied and Environmental Microbiology*, 71(7): 3504-3511.

Kujawska, A., Kujawski, J., Bryjak, M. and Kujawski, W. (2015). ABE fermentation products recovery methods—a review. *Renewable and Sustainable Energy Reviews*, 48: 648-661.

Kurkijärvi, A., Lehtonen, J. and Linnekoski, J. (2014). Novel dual extraction process for acetone–butanol–ethanol fermentation. *Separation and Purification Technology*, 124:18–25.

Lee, S. H. and Eom, M. H. (2018). *Recombinant microorganisms with improved butanol production ability and method for producing butanol by using the same*. U.S. Patent No. 9,957,529 B2.

Lee, S. Y., Park, J. H., Jang, S. H., Nielsen, L. K., Kim, J. and Jung, K. S. (2008). Fermentative butanol production by Clostridia. *Biotechnology and Bioengineering*, 101(2): 209-228.

Li, Y., Meng, L., Nithyanandan, K., Lee, T. H., Lin, Y., Lee, C. F. and Liao, S. (2016). Combustion, performance and emissions characteristics of a spark-ignition engine fueled with isopropanol-n-butanol-ethanol and gasoline blends. *Fuel*, 184: 864-872.

Li, J., Shi, S., Adhikari, S., Tu, M. (2017). Inhibition effect of aromatic aldehydes on butanol fermentation by Clostridium acetobutylicum. *Royal Society of Chemistry Advances*. 7:1241-1250.

Lienhardt, J., Schripsema, J., Qureshi, N. and Blaschek, H. P. (2002). Butanol production by Clostridium beijerinckii BA101 in an immobilized cell biofilm reactor. *Applied Biochemistry and Biotechnology*, 98(1-9): 591-598.

Lin, X, Wu, J., Fan, J., Qian, W., Zhou, X., Qian, C., Jin, X., Wang, L., Bai, J. and Ying, H. (2012). Adsorption of butanol from aqueous solution onto a new type of macroporous adsorption resin: studies of adsorption isotherms and kinetics simulation. *Journal of Chemical Technology and Biotechnology*, 87: 924–931.

Liu, G., Gan, L., Liu, S., Zhou, H., Wei, W. and Jin, W. (2014). PDMS/ceramic composite membrane for pervaporation separation of acetone-butanol-ethanol (ABE) aqueous solutions and its application in intensification of ABE fermentation process. *Chemical Engineering and Processing*, 86: 162-172.

Loneylinked, Q. (2017). *Cathay Dupont Award: Cathay Industrial Biotech files for IPO*. At https://medium.com/@loneyqeturah/cathay-dupont-award-cathay-industrial-biotech-files-for-ipo-bef957d277f. Last visited June 24, 2020.

López-Contreras, A. M., Claassen, P. A., Mooibroek, H. and De Vos, W. M. (2000). Utilisation of saccharides in extruded domestic organic waste by Clostridium acetobutylicum ATCC 824 for production of acetone, butanol and ethanol. *Applied Microbiology and Biotechnology,* 54(2): 162-167.

Lu, C., Dong, J., Yang, S. T. (2013). Butanol production from wood pulping hydrolysate in an integrated fermentation-gas stripping process. *Bioresource Technology*. 143: 467-475.

Luo, H., Zheng, P., Bilal, M., Xie, F., Zeng, Q., Zhu, C., Yang, R., Wang, Z. (2020). Efficient bio-butanol production from lignocellulosic waste by elucidating the mechanisms of Clostridium acetobutylicum response to phenolic inhibitors. *Science of The Total Environment*, 710: 136399.

Luyben, W. L. (2012). Pressure-swing distillation for minimum- and maximum- boiling homogeneous azeotropes. *Industrial & Enginering Chemistry Research*, 51(33):10881–10886.

Marchal, R., Blanchet, D. and Vandecasteele, J. P. (1985). Industrial optimization of acetone-butanol fermentation: a study of the utilization of Jerusalem artichokes. *Applied Microbiology and Biotechnology*, 23(2): 92-98.

Maddox, I. S., Qureshi, N. and Gutierrez, N. A. (1993). Utilization of whey by clostridia and process technology. In: Woods, D.R. (Ed.), *The Clostridia and Biotechnology*. Butterworth-Heinemann: 343-369.

Marchal, R., Ropars, M. and Vandecasteele, J. P. (1986). Conversion into acetone and butanol of lignocellulosic substrates pretreated by steam explosion. *Biotechnology Letters*, 8(5): 365-370.

Mariano, A. P., Dias, M. O. S., Junqueira, T. L., Cunha, M. P., Bonomi, A. and Maciel Filho, R. (2013). Butanol production in a first-generation Brazilian sugarcane biorefinery: Technical aspects and economics of greenfield projects. *Bioresource Technology*, 135: 316-323.

Mariano, A. P., Keshtkar, M. J., Atala, D. I. P., Maugeri Filho, F., Wolf Maciel M. R., Maciel Filho, R., Stuart, P. (2011). Energy requirements for butanol recovery using the flash fermentation technology. *Energy & Fuels*, 25(5):2347–2355.

Marzo, C., Díaz, A. B., Caro, I. and Blandino, A. (2019). Status and perspectives in bioethanol production from sugar beet. In: Ray, R., Ramachandran, S. (Eds.), *Bioethanol production from food crops*. Academic Press, Cambridge: 61-79.

Masngut, N. and Harvey, A. P. (2012). Intensification of biobutanol production in batch oscillatory baffled bioreactor. *Procedia Engineering*, 42: 1079-1087.

Matsumura, M., Kataoka, H., Sueki, M., Araki, K. (1988). Energy saving effect of pervaporation using oleyl alcohol liquid membrane in butanol purification. *Bioprocess Engineering*, 3:93–100.

Miah, J. H., Griffiths, A., McNeill, R., Halvorson, S., Schenker, U., Espinosa-Orias, N. D., Morse, S., Yang, A. and Sadhukhan, J. (2018). Environmental management of confectionery products: life cycle impacts and improvements strategies. *Journal of Cleaner Production*, 177: 732-751.

Mohanty, S. K. and Swain, M. R. (2019). Bioethanol production from corn and wheat: food, fuel, and future. In: Ray, R., Ramachandran, S. (Eds.), *Bioethanol production from food crops*. Academic Press, Cambridge: 45-79.

Moreno, J. A. and Cubillos Lobo, J. A. (2017). Bioetanol como combustible: una alternativa sustentable. *Investigación Joven*, 4(1): 45-50 (Spanish).

Moretti, M. M. de S., Bocchini-Martins, D. A., Nunes, C. da C. C., Villena, M. A., Perrone, O. M., Silva, R. da, Boscolo, M., Gomes, E. (2014). Pretreatment of sugarcane bagasse with microwaves irradiation and its effects on the structure and on enzymatic hydrolysis. *Applied Energy*, 122: 189-195.

Moriarty, K., Milbrandt, A., Lewis, J. and Schwab, A. (2020). *2017 Bioenergy industry status report*. Technical Report NREL/TP-5400-75776. National Renewable Energy Laboratory.

Nakas, J. P., Schaedle, M., Parkinson, C. M., Coonley, C. E. and Tanenbaum, S. W. (1983). System development for linked-fermentation production of solvents from algal biomass. *Applied and Environmental Microbiology*, 46(5): 1017-1023.

Ni, Y. and Sun, Z. (2009). Recent progress on industrial fermentative production of acetone-butanol-ethanol by *Clostridium acetobutylicum* in China. *Applied Microbiology and Biotechnology*, 83: 415-423.

Ni, Y., Xia, Z., Wang, Y., Sun, Z. (2013). Continuous butanol fermentation from inexpensive sugar-based feedstocks by Clostridium saccharobutylicum DSM 13864. *Bioresource Technology*, 129: 680-685.

Nielsen, D. R., Leonard, E., Yoon, S. H., Tseng, H. C., Yuan, C. and Prather, K. L. J. (2009). Engineering alternative butanol production platforms in heterologous bacteria. *Metabolic engineering*, 11(4-5): 262-273.

Nguyen, N. P. T., Raynaud, C., Meynial-Salles, I. and Soucaille, P. (2018). Reviving the Weizmann process for commercial n-butanol production. *Nature Communications*, 9: 3682.

Nölling, J., Breton, G., Omelchenko, M. V., Makarova, K. S., Zeng, Q., Gibson, R., Lee, H. M., Dubois, J., Qiu, D., Hitti, J., Wolf, Y. I., Tatusov, R. L., Sabathe, F., Doucette-Stamm, L., Soucaille, P., Daly, M. J., Bennett, G. N., Koonin, E. V. and Smith, D. R. (2001). Genome sequence and comparative analysis of the solvent-producing bacterium Clostridium acetobutylicum. *Journal of Bacteriology*, 183(16): 4823-4838.

Oudshoorn, A., van der Wielen, L. A. M. and Straathof A. J. J. Adsorption equilibria of bio-based butanol solutions using zeolite. *Biochemical Engineering Journal*, 48(1):99–103.

Park, C. H., Okos, M. R. and Wankat, P. C. (1991). Acetone–butanol–ethanol (ABE) fermentation and simultaneous separation in a trickle bed reactor. *Biotechnology Progress*, 7(2): 185–194.

Pérez, J., Muñoz-Dorado, J., De La Rubia, T., Martínez, J. (2002). Biodegradation and biological treatments of cellulose, hemicellulose and lignin: An overview. *International Microbiology*, 5: 53-63.

Qureshi, N. and Blaschek, H. P. (2000). Butanol production using Clostridium beijerinckii BA101 hyper-butanol producing mutant strain and recovery by pervaporation. In: Finkelstein, M., Davidson, B.H. (Eds.), *Twenty-First Symposium on Biotechnology for Fuels and Chemicals*. Humana Press, New Jersey: 225-235.

Qureshi, N. and Blaschek, H. P. (2005). Butanol production from agricultural biomass. In: Shetty, K., Paliyath, G., Pometto, A., Levin, R.E. (Eds.), *Food Biotechnology*. Taylor & Francis, Boca Raton: 525-549.

Qureshi, N. and Blaschek, H. P. (2001a). ABE production from corn: a recent economic evaluation. *Journal of Industrial Microbiology and Biotechnology*, 27(5): 292-297.

Qureshi, N. and Blaschek, H. (2001b). Evaluation of recent advances in butanol fermentation, upstream, and downstream processing. *Bioprocess and Biosystems Engineering*, 24(4): 219-226.

Qureshi, N., Bowman, M. J., Saha, B. C., Hector, R., Berhow, M. A., Cotta, M. A. (2012). Effect of cellulosic sugar degradation products (furfural and hydroxymethyl furfural) on acetone-butanol-ethanol (ABE) fermentation using Clostridium beijerinckii P260. *Food and Bioproducts Processing*, 90: 533-540.

Qureshi, N., Ezeji, T. C., Ebener, J., Dien, B. S., Cotta, M. A. and Blaschek, H. P. (2008). Butanol production by Clostridium beijerinckii. Part I: use of acid and enzyme hydrolyzed corn fiber. *Bioresource Technology*, 99(13): 5915-5922.

Qureshi, N., Lolas, A. and Blaschek, H. P. (2001). Soy molasses as fermentation substrate for production of butanol using Clostridium beijerinckii BA101. *Journal of Industrial Microbiology and Biotechnology*, 26(5): 290-295.

Qureshi, N. and Maddox, I. S. (2005). Reduction in butanol inhibition by perstraction: utilization of concentrated lactose/whey permeate by Clostridium acetobutylicum to enhance butanol fermentation economics. *Food and Bioproducts Processing*, 83(1): 43-52.

Qureshi, N., Meagher, M. M. and Hutkins, R. W. (1999). Recovery of butanol from model solutions and fermentation broth using a silicalite/silicone membrane. *Journal of Membrane Science*, 158(1-2): 115-125.

Qureshi, N., Saha, B. C., Dien, B., Hector, R. E. and Cotta, M. A. (2010). Production of butanol (a biofuel) from agricultural residues: Part I–Use of barley straw hydrolysate. *Biomass and Bioenergy*, 34(4): 559-565.

Qureshi, N., Schripsema, J., Lienhardt, J. and Blaschek, H. P. (2000). Continuous solvent production by Clostridium beijerinckii BA101 immobilized by adsorption onto brick. *World Journal of Microbiology and Biotechnology*, 16(4): 377-382.

Rabemanolontsoa, H., Saka, S. (2016). Various pretreatments of lignocellulosics. *Bioresource Technology*, 199: 83-91.

Ramey, D. E. (1998). *Continuous two stage, dual path anaerobic fermentation of butanol and other organic solvents using two different strains of bacteria*. U.S. Patent No. 5,753,474.

Raganati, F., Procentese, A., Olivieri, G., Russo, M. E., Salatino, P. and Marzocchella, A. (2020). Bio-butanol recovery by adsorption/desorption processes. *Separation and Purification Technology*, 235, 116145.

Reyes, L. H., Almario, M. P. and Kao, K. C. (2011). Genomic library screens for genes involved in n-butanol tolerance in Escherichia coli. *PloS ONE*, 6(3): e17678.

Richardson, J. F., Harker, J. H. and Backhurst, J. R. (2002). Product design and process intensification. In: Coulson, J. M., Richardson, J. F., Backhurst, J. R. Harker, J. H., (Eds.), *Chemical Engineering (Fifth Edition) Volume 2: Particle Technology and Separation Process*. Butterworth-Heinemann: 1104-1135.

Rios-González, L. J., Morales-Martínez, T. K., Rodríguez-Flores, M. F., Rodríguez-De la Garza, J. A., Castillo-Quiroz, D., Castro-Montoya, A.J. and Martinez, A. (2017). Autohydrolysis pretreatment assessment in ethanol production from agave bagasse. *Bioresource Technology*, 242: 184-190.

Rodriguez-Gomez, D., Lehmann, L., Schultz-Jensen, N., Bjerre, A. B., Hobley, T. J. (2012). Examining the potential of plasma-assisted pretreated wheat straw for enzyme production by *Trichoderma reesei*. *Applied Biochemistry and Biotechnology*. 166: 2051–2063.

Rom, A., Wukovits, W. and Anton, F. (2014). Development of a vacuum membrane distillation unit operation: from experimental data to a simulation model. *Chemical Engineering and Processing: Process Intensification*, 86: 90–95.

Romero-García, A. G., Prado-Rúbio, O. A., Contreras-Zarazúa, G., Ramírez-Márquez, C., Segovia-Hernández, J. G. (2019). Simultaneous design and controllability optimization for the reaction zone for furfural bioproduction, *Computer Aided Chemical Engineering*. 46: 133-138.

Rühl, J., Schmid, A. and Blank, L. M. (2009). Selected Pseudomonas putida strains able to grow in the presence of high butanol concentrations. *Applied and Environtal Microbiology*, 75(13): 4653-4656.

Patraşcu I., Bildea, C. S. and Kiss, A. A. (2018). Eco-efficient downstream processing of biobutanol by enhanced process intensification and integration. *ACS Sustainable Chemistry & Engineering*, 6(4): 5452-5461.

Patraşcu, I., Bildea, C. S. and Kiss, A. A. (2019). Dynamics and control of a heat pump assisted azeotropic dividing-wall column for biobutanol purification. *Chemical Engineering Research and Design*, 146, 416-426.

Qian Z., Chen, Q. and Grossmann, I. E. (2018). Optimal synthesis of rotating packed bed and packed bed: a case illustrating the integration of PI and PSE. *Computer Aided Chemical Engineering*, 44: 2377-2382.

Qureshi, N., Hughes, S., Maddox, I. S., and Cotta, M. A. (2005). Energy-efficient recovery of butanol from model solutions and fermentation broth by adsorption, *Bioprocess and Biosystems Engineering*, 27(4): 215–222.

Ramos J. L., Valdivia, M., García-Lorente, F. and Segura, A. (2016). Benefits and perspectives on the use of biofuels. *Microbial Biotechnology*, 9(4): 436-440.

Rose, A. H. (1961). *Industrial Microbiology*. Butterworths, London.

Ruggeri, B., Tommasi, T. and Sanfilippo, S. (2015). *BioH$_2$ & BioCH$_4$ Through Anaerobic Digestion: From Research to Full-Scale Application*, Springer, London.

Sadegh, H. and Ali, G. A. (2019). Potential applications of nanomaterials in wastewater treatment: nanoadsorbents performance. In: Hussain, A., Ahmed, S. (Eds.), *Advanced Treatment Techniques for Industrial Wastewater*. IGI Global: 51-61.

Saleh, T. A., & Gupta, V. K. (2016). An overview of membrane science and technology. In: Saleh, T. A., Gupta, V. K. (Eds.) *Nanomaterial and Polymer Membranes*. Elsevier, Amsterdam: 1-23.

Sanchez-Orozco, R., Balderas Hernández, P., Roa Morales, G., Ureña Núñez, F., Orozco Villafuerte, J., Lugo Lugo, V., Flores Ramírez, N., Barrera Díaz, C. E. and Cajero Vázquez, P. (2014). Characterization of lignocellulosic fruit waste as an alternative feedstock for bioethanol production. *Bioresources*, 9(2): 1873-1885.

Sánchez-Ramírez, E., Quiroz-Ramírez, J. J., Segovia-Hernández, J. G., Hernández, S. and Bonilla-Petriciolet, A. (2015). Process alternative for biobutanol purification: design and optimization. *Industrial and Engineering Chemistry Research*, 54(1): 351-358.

Sánchez-Ramírez, E., Quiroz-Ramírez, J. J., Hernández, S., Segovia-Hernández, J. G., and Kiss, A. A. (2017). Optimal hybrid separations for intensified downstream processing of biobutanol. *Separation and Purification Technology*, 185: 149-159.

Sánchez-Ramírez, E., Ponce-Rocha, J. D., Segovia-Hernández, J. G., Gómez-Castro, F. I. and Morales-Rodríguez, R. (2018). A framework for optimised sustainable solvent mixture and separation process design. *Computed Aided Chemical Engineering*, 43: 755-760.

Satlewal, A., Agrawal, R., Bhagia, S., Sangoro, J., Ragauskas, A. J. (2018). Natural deep eutectic solvents for lignocellulosic biomass pretreatment: Recent developments, challenges and novel opportunities. *Biotechnology Advances*. 36: 2032-2050.

Saucedo-Luna, J., Castro-Montoya, A. J., Martínez-Pacheco, M. M., Sosa-Aguirre, C. R. and Campos-Garcia, J. (2010). Efficient chemical and enzymatic saccharification of the lignocellulosic residue from *Agave tequilana* bagasse to produce ethanol by *Pichia caribbica*. *Journal of Industrial Microbiology and Biotechnology*, 38: 725-732.

Segovia-Hernández, J. G., Sánchez-Ramírez, E., Alcocer-García, H., Quiroz-Ramírez, J. J., Udugama, I. A. and Mansouri, S. S. (2020). Analysis of intensified sustainable schemes for biobutanol purification. *Chemical Engineering and Processing*, 147: 107737.

Setlhaku, M., Heitmann, S., Gorak, A. and Wichmann, R. (2013) Investigation of gas stripping and pervaporation for improved feasibility of two-stage butanol production process. *Bioresource Technology*, 136:102–108.

Shaheen, R., Shirley, M. and Jones, D. T. (2000). Comparative fermentation studies of industrial strains belonging to four species of solvent-producing clostridia. *Journal of Molecular Microbiology and Biotechnology*, 2(1): 115-124.

Sharma, H. K., Xu, C., Qin, W., 2019. Biological Pretreatment of Lignocellulosic Biomass for Biofuels and Bioproducts: An Overview. *Waste and Biomass Valorization*. 10: 235-251.

Sharma, P. and Chung, W. J. (2011) Synthesis of MEL type zeolite with different kinds of morphology for the recovery of 1-butanol from aqueous solution. *Desalination*, 275(1-3):172–180.

Shen, C. R., Lan, E. I., Dekishima, Y., Baez, A., Cho, K. M. and Liao, J. C. (2011). Driving forces enable high-titer anaerobic 1-butanol synthesis in Escherichia coli. *Applied and Environmental Microbiology*, 77(9): 2905-2915.

Sindhu, R., Binod, P., Pandey, A. (2016). Biological pretreatment of lignocellulosic biomass - An overview. *Bioresource Technology*, 199: 76-82.

Siegmeier, T., Blumenstein, B. and Möller, D. (2019). Bioenergy production and organic agriculture. In: Chandran, S., Unni, M. R., Thomas, S. (Eds.), *Organic farming: Global perspectives and methods*. Woodhead Publishing, Duxford: 331-359.

Silalertruksa, T. and Gheewala, S. (2018). Land-water-energy nexus of sugarcane production in Thailand. *Journal of Cleaner Production*, 182: 521-528.

Schnurbusch, T. (2019). Wheat and barley biology: towards new frontiers. *Journal of Integrative Plant Biology*, 61: 198-203.

Smith, R. and Jobson, M. (2000). Distillation. In: Wilson, I.D. (Ed.), *Encyclopedia of Separation Science*. Academic Press: 84-103.

Steen, E. J., Chan, R., Prasad, N., Myers, S., Petzold, C. J., Redding, A., Ouellet, M. and Keasling, J. D. (2008). Metabolic engineering of Saccharomyces cerevisiae for the production of n-butanol. *Microbial Cell Factories*, 7(1): 36.

Sun, S., Sun, S., Cao, X., Sun, R. (2016). The role of pretreatment in improving the enzymatic hydrolysis of lignocellulosic materials. *Bioresource Technology*, 199: 49-58.

Survase, S. A., Sklavounos, E., Jurgens, G., van Heiningen, A., Granström, T. (2011). Continuous acetone-butanol-ethanol fermentation using SO 2-ethanol-water spent liquor from spruce. *Bioresource Technology*, 102(23):10996-11002.

Syed, Q. (1994). *Biochemical studies on anaerobic fermentation of molasses by Clostridium acetobutylicum*. PhD dissertation, University of Punjab.

Tashiro, Y., Takeda, K., Kobayashi, G. and Sonomoto, K. (2005). High production of acetone–butanol–ethanol with high cell density culture by cell-recycling and bleeding. *Journal of Biotechnology*, 120(2): 197-206.

Tilman, D., Hill, J. and Lehman, C. (2006), Carbon-Negative Biofuels from Low-Input High-Diversity Grassland Biomass, *Science*, 314: 1598-1600.

Tiwari, S., Jadhav, S. K., Sharma, M. and Tiwari, K. L. (2014). Fermentation of waste fruits for bioethanol production. *Asian Journal of Biological Sciences*, 7(1): 30-34.

U.S. Department of Energy (n.d.). *Alternative fuels data center.* At https://afdc.energy.gov/fuels/emerging_biobutanol.html. Last visited April 4th, 2020.

van der Merwe, A. B., Cheng, H., Görgens, J. F., Knoetze, J. H. (2013) Comparison of energy efficiency and economics of process designs for biobutanol production from sugarcane molasses. *Fuel*, 105: 451-458.

Velázquez-Valadez, U., Farías-Sánchez, J. C., Vargas-Santillán, A. and Castro-Montoya, A. J. (2016). *Tequilana weber* agave bagasse enzymatic hydrolysis for the production of fermentable sugars: oxidative-alkaline pretreatment and kinetic modeling. *BioEnergy Research*, 9: 998-1004.

Voget, C. E., Mignone, C. F. and Ertola, R. J. (1985). Butanol production from apple pomace. *Biotechnology Letters*, 7(1): 43-46.

Wang, L., Chen, H. (2011). Increased fermentability of enzymatically hydrolyzed steam-exploded corn stover for butanol production by removal of fermentation inhibitors. *Process Biochemistry*, 46: 604-607.

Wang, M., Huo, H. and Arora, S. (2011a). Methods of dealing with co-product of biofuels in life-cycle analysis and consequent results within the U.S. context. *Energy Policy,* 39(10): 5726-5736.

Wang, Y., Ho, S. H., Yen, H. W., Nagarajan, D., Ren, N. Q., Li, S., Hu, Z., Lee, D. J., Kondo, A. and Chang, J. S. (2017). Current advances on fermentative biobutanol production using third generation feedstock. *Biotechnology Advances,* 35(8): 1049-1059.

Wang, S., Zhang, Y., Dong, H., Mao, S., Zhu, Y., Wang, R., Luan, G., Li, Y. (2011b). Formic acid triggers the "acid crash" of acetone-butanol-ethanol fermentation by Clostridium acetobutylicum. *Applied and Environmental Microbiology*, 77(5): 1674-1680.

Wiehn, M., Staggs, K., Wang, Y. C. and Nielsen D. R. (2014). In situ butanol recovery from Clostridium acetobutylicum fermentations by expanded bed adsorption. *Biotechnology Progress*, 30(1):68–78.

Winkler, J. and Kao, K. C. (2011). Transcriptional analysis of Lactobacillus brevis to N-butanol and ferulic acid stress responses. *PLoS ONE*, 6(8): e21438.

Winkler, J., Rehmann, M. and Kao, K. C. (2010). Novel Escherichia coli hybrids with enhanced butanol tolerance. *Biotechnology Letters*, 32(7): 915-920.

Woods, D. R. (1995). The genetic engineering of microbial solvent production. *Trends in Biotechnology*, 13(7): 259-264.

Wu, C. and Tu, X. (2016). Biological and fermentative conversion of syngas. In: Luque, R., Lin, C. S. K., Wilson, K, Clark, J. (Eds.), *Handbook of Biofuels Production*. Woodhead Publishing, Duxford: 335-357.

Xing, W., Xu, G., Dong, J., Han, R., Ni, Y. (2018). Novel dihydrogen-bonding deep eutectic solvents: Pretreatment of rice straw for butanol fermentation featuring enzyme recycling and high solvent yield. *Chemical Engineering Journal*, 333: 712-720.

Yang, X. and Tsao, G. T. (1995). Enhanced acetone-butanol fermentation using repeated fed-batch operation coupled with cell recycle by membrane and simultaneous removal of inhibitory products by adsorption. *Biotechnology and Bioengineering*, 47(4): 444-450.

Zhang, Y., Ma, Y., Yang, F. and Zhang, C. (2009). Continuous acetone–butanol–ethanol production by corn stalk immobilized cells. *Journal of Industrial Microbiology & Biotechnology*, 36(8): 1117-1121.

Zhang, K., Pei, Z., Wang, D. (2016). Organic solvent pretreatment of lignocellulosic biomass for biofuels and biochemicals: A review. *Bioresource Technology*, 199: 21-33.

Zhang, J., Wang, M., Gao, M., Fang, X., Yano, S., Qin, S., Xia, R. (2013). Efficient Acetone-Butanol-Ethanol Production from Corncob with a New Pretreatment Technology-Wet Disk Milling. *Bio Energy Research*, 6: 35-43.

Zverlov, V. V., Berezina, O., Velikodvorskaya, G. A., and Schwarz, W. H. (2006). Bacterial acetone and butanol production by industrial fermentation in the Soviet Union: use of hydrolyzed agricultural waste for biorefinery. *Applied Microbiology and Biotechnology*, 71(5): 587-597.

BIOGRAPHICAL SKETCHES

Fernando Israel Gómez Castro

Affiliation: Departamento de Ingeniería Química, División de Ciencias Naturales y Exactas, Campus Guanajuato, Universidad de Guanajuato

Education: ScD in Chemical Engineering

Business Address: Noria Alta S/N Col. Noria Alta, Guanajuato, Gto., México, 36050

Research and Professional Experience: Researcher in the area of process design, analysis and optimization, recognized as National Researcher by the National System of Researchers (SNI, México). 37 scientific papers published in indexed journals, 8 book chapters and 4 books. Participation in several national and international scientific conferences. Reviewer for different indexed journals. Current Head of the Bachelor Program on Chemical Engineering of the Universidad de Guanajuato.

Professional Appointments: Member of the Directive Board of the Mexican Academy of Research and Teaching in Chemical Engineering (AMIDIQ). Member of the American Chemical Society. Registered as accredited evaluator for the National Council of Science and Technology (Mexico). Member of the Thematic Network on Bioenergy (Mexico). Member of the Editorial Board of the *Journal of Energy, Engineering Optimization and Sustainability*.

Honors: Elected as Member of the Directive Board of the Mexican Academy of Research and Teaching in Chemical Engineering (AMIDIQ) in the periods 2017-2019 and 2019-2021. 2017 Certificate of Excellence in Reviewing (Clean Technologies and Environmental Policy). 2018 Certificate of Outstanding Contribution in Reviewing (Applied Soft Computing).

Publications from the Last 3 Years:

- May-Vázquez M. M., Rodríguez-Ángeles M. A., Gómez-Castro F. I., Espinoza-Zamora J., Murrieta-Luna E., 2020, Development of a mass transfer model for the rate-based simulation of a batch distillation column, *Computers & Chemical Engineering*, 140, 106981.
- Gutiérrez-Antonio C., Romero-Izquierdo A. G., Gómez-Castro F. I., Hernández S., *"Production Processes for Renewable Aviation Fuel: Present Technologies and Future Trends,"* Elsevier, 2020 (ISBN-13: 978-0128197196).
- López-Molina A., Gómez-Castro F. I., 2020, Non-catalytic production of biodiesel: energy and safety considerations, en *Biofuels: Advances in research and applications*, Edited by G.R. Carey, Nova Publishers (ISBN: 978-1-53617-721-3), pp. 1-29.
- Gómez-Castro F. I., Segovia-Hernández J. G. (Eds.), *"Process Intensification: Design Methodologies,"* De Gruyter, 2019 (ISBN-10: 3110596075, ISBN-13: 978-3110596076).
- Contreras-Vargas C. A., Gómez-Castro F. I., Sánchez-Ramírez E., Segovia-Hernández J. G., Morales-Rodríguez R., Gamiño-Arroyo Z., 2019, Alternatives for the purification of the blend butanol/ethanol from an ABE fermentation effluent: impact on the economic, environmental and safety indexes, *Chemical Engineering & Technology*, 42, 1088-1100.
- Arce-Alejandro R., Villegas-Alcaraz J. F., Gómez-Castro F. I., Juárez-Trujillo L., Sánchez-Ramírez E., Carrera-Rodríguez M., Morales-Rodríguez R., 2018, Performance of a gasoline engine

powered by a mixture of ethanol and n-butanol, *Clean Technologies and Environmental Policy,* 20, 1929-1937.
- Romero-Izquierdo A. G., Gutiérrez-Antonio C., Gómez-Castro F. I., Hernández S., 2018, Hydrotreating of triglyceride feedstock to produce renewable aviation fuel, *Recent Innovations in Chemical Engineering,* 11, 77-89.
- Segovia-Hernández J. G., Gómez-Castro F. I., Sánchez-Ramírez E., 2018, Dynamic performance of a complex distillation configuration for the separation of a five-components hydrocarbon mixture, *Chemical Engineering & Technology,* 41, 2053-2065.
- Gutiérrez-Antonio C., Gómez-De la Cruz A., Romero-Izquierdo A. G., Gómez-Castro F. I., Hernández S., 2018, Modeling, simulation and intensification of hydroprocessing of micro-algae oil to produce renewable aviation fuel, *Clean Technologies and Environmental Policy,* 20, 1589-1598.
- Alfaro-Ayala J. A., López-Núñez O. A., Gómez-Castro F. I., Ramírez-Minguela J. J., Uribe-Ramírez A. R., Belman-Flores J. M., Cano-Andrade S., 2018, Optimization of a solar collector with evacuated tubes using the simulated annealing and computational fluid dynamics, *Energy Conversion and Management,* 166, 343-355.
- Velázquez-Guevara M.A., Uribe-Ramírez A. R., Gómez-Castro F. I., Ponce-Ortega J. M., Hernández S., Segovia-Hernández J. G., Alfaro-Ayala J. A., Ramírez-Minguela J. J., 2018, Synthesis of mass exchange networks: a novel mathematical programming approach, *Computers and Chemical Engineering,* 115, 226-232.
- Gutiérrez-Antonio C., Soria-Ornelas M. L., Gómez-Castro F. I., Hernández S., 2018, Intensification of the hydrotreating process to produce renewable aviation fuel through reactive distillation, *Chemical Engineering and Processing: Process Intensification,* 124, 122-130.

Eduardo Sánchez-Ramírez

Affiliation: Departamento de Ingeniería Química, División de Ciencias Naturales y Exactas, Campus Guanajuato, Universidad de Guanajuato

Education: ScD in Chemical Engineering

Business Address: Noria Alta S/N Col. Noria Alta, Guanajuato, Gto., México, 36050

Research and Professional Experience: Professor at the Department of Chemical Engineering at the University of Guanajuato (Mexico) since 2017. Through his academic development, he has gained considerable experience in the area of synthesis, design, simulation, control and optimization of chemical processes. Currently published contributions focus on the production of biofuels and base chemicals in the chemical industry. He has currently published more than 25 articles in indexed journals, 6 book chapters from renowned publishers and has registered 2 patents. He acts as a reviewer of indexed journals in the area of energy and chemical engineering.

Awards and Distinctions: Summa Cum Laude Ph.D. Degree Holder (March, 2017). National System of Researchers Level 1, SNI Mexico, (January, 2019). Guest Editor at *Chemical Engineering and Processing Process Intensification Journal* May, 2020.

Publications from the Last 3 Years:

- Sánchez-Ramírez, E., Alcocer-García, H., Quiroz-Ramírez, J. J., Ramírez-Márquez, C., Segovia-Hernández, J. G., Hernández, S., Errico, M., Castro-Montoya, A. J., 2017, Control Properties of Hybrid Distillation Processes for the Separation of Biobutanol, *Journal of Chemical Technology & Biotechnology*, 92, 959 – 970.

- Quiroz-Ramírez, J. J., Sánchez-Ramírez, E., Hernández-Castro, S., Ramírez-Prado, J. H., Segovia-Hernández, J. G., 2017, Multi-Objective Stochastic Optimization Approach Applied to a Hybrid Process Production-Separation in the Production of Biobutanol, *Ind. Eng. Chem. Res.*, 56, 1823 - 1833.
- Quiroz-Ramírez, J. J., Sánchez-Ramírez, E., Segovia-Hernández, J. G., Hernández S., Ponce-Ortega, J. M., 2017, Optimal Selection of Feedstock for Biobutanol Production Considering Economic and Environmental Aspects, *ACS Sustainable Chemistry & Engineering*, 5, 4018 − 4030.
- Sánchez-Ramírez, E., Quiroz-Ramírez, J. J., Hernández S, Segovia-Hernández, J. G., Kiss, A. A., 2017, Optimal Hybrid Separations for Intensified Downstream Processing of Biobutanol, *Separation and Purification Technology,* 185, 149 - 159.
- Errico, M., Sanchez-Ramirez, E., Quiroz-Ramìrez, J. J., Rong, B. G., & Segovia-Hernandez, J. G. (2017). Multiobjective Optimal Acetone–Butanol–Ethanol Separation Systems Using Liquid–Liquid Extraction-Assisted Divided Wall Columns. *Industrial & Engineering Chemistry Research,* 56(40), 11575-11583.
- Quiroz-Ramírez, J. J., Sánchez-Ramírez, E., & Segovia-Hernández, J. G. (2018). Energy, exergy and techno-economic analysis for biobutanol production: a multi-objective optimization approach based on economic and environmental criteria. *Clean Technologies and Environmental Policy,* 1-22.
- Sánchez-Ramírez, E., Ramírez-Márquez, C., Quiroz-Ramírez, J. J., Contreras-Zarazúa, G., Segovia-Hernández, J. G., & Cervantes-Jauregui, J. A. (2018). Reactive Distillation Column Design for Tetraethoxysilane (TEOS) Production: Economic and Environmental Aspects. *Industrial & Engineering Chemistry Research.* 2018, 57, 5024−5034.
- Torres-Ortega, C. E., Ramírez-Márquez, C., Sánchez-Ramírez, E., Quiroz-Ramírez, J. J., Segovia-Hernandez, J. G., & Rong, B. G. (2018). Effects of intensification on process features and control properties of lignocellulosic bioethanol separation and dehydration

- systems. *Chemical Engineering and Processing-Process Intensification.*
- Arce-Alejandro, R., Villegas-Alcaraz, J. F., Gómez-Castro, F. I., Juárez-Trujillo, L., Sánchez-Ramírez, E., Carrera-Rodríguez, M., & Morales-Rodríguez, R. (2018). Performance of a gasoline engine powered by a mixture of ethanol and n-butanol. *Clean Technologies and Environmental Policy,* 1-9.
- Hernández, J. G. S., Gómez-Castro, F. I., & Sánchez-Ramírez, E. Dynamic Performance of a Complex Distillation Configuration for the Separation of a Five-Components Hydrocarbon Mixture. *Chemical Engineering & Technology.*
- Sánchez-Ramírez, E., Quiroz-Ramírez, J. J., Hernández, S., Hernández, J. G. S., Contreras-Zarazúa, G., & Ramírez-Márquez, C. (2019). Synthesis, design and optimization of alternatives to purify 2, 3-Butanediol considering economic, environmental and safety issues. *Sustainable Production and Consumption,* 17, 282-295.
- Contreras-Zarazúa, G., Sánchez-Ramírez, E., Vazquez-Castillo, J. A., Ponce-Ortega, J. M., Errico, M., Kiss, A. A., & Segovia-Hernández, J. G. (2018). Inherently safer design and optimization of intensified separation processes for furfural production. *Industrial & Engineering Chemistry Research,* 58(15), 6105-6120.
- Contreras-Vargas, C. A., Gómez-Castro, F. I., Sánchez-Ramírez, E., Segovia-Hernández, J. G., Morales-Rodríguez, R., & Gamiño-Arroyo, Z. (2019). Alternatives for the Purification of the Blend Butanol/Ethanol from an Acetone/Butanol/Ethanol Fermentation Effluent. *Chemical Engineering & Technology,* 42(5), 1088-1100.
- Alcocer-García, H., Segovia-Hernández, J. G., Prado-Rubio, O. A., Sánchez-Ramírez, E., & Quiroz-Ramírez, J. J. (2020). Multi-objective optimization of intensified processes for the purification of levulinic acid involving economic and environmental objectives. Part II: A comparative study of dynamic properties. *Chemical Engineering and Processing-Process Intensification,* 147, 107745.
- Alcocer-García, H., Segovia-Hernández, J. G., Prado-Rubio, O. A., Sánchez-Ramírez, E., & Quiroz-Ramírez, J. J. (2020). Multi-

objective optimization of intensified processes for the purification of levulinic acid involving economic and environmental objectives. Part II: A comparative study of dynamic properties. *Chemical Engineering and Processing-Process Intensification*, 147, 107745.
- Segovia-Hernández, J. G., Sánchez-Ramírez, E., Alcocer-García, H., Quíroz-Ramírez, J. J., Udugama, I. A., & Mansouri, S. S. (2020). Analysis of intensified sustainable schemes for biobutanol purification. *Chemical Engineering and Processing-Process Intensification*, 147, 107737.
- Sánchez-Ramírez, E., Ramírez-Márquez, C., Quiroz-Ramírez, J. J., Angelina-Martínez, A. Y., Cortazar, V. V., & Segovia-Hernández, J. G. (2020). Design of dividing wall columns involving sustainable indexes for a class of quaternary mixtures. *Chemical Engineering and Processing-Process Intensification*, 148, 107833.
- Torres-Vinces, L., Contraras-Zarazua, G., Huerta-Rosas, B., Sánchez-Ramírez, E., & Segovia-Hernández, J. G. Methyl-Ethyl Ketone Production through Novel Intensified Process. *Chemical Engineering & Technology*.

Ricardo Morales-Rodriguez

Affiliation: Departamento de Ingeniería Química, División de Ciencias Naturales y Exactas, Campus Guanajuato, Universidad de Guanajuato

Education: PhD in Chemical Engineering

Business Address: Noria Alta S/N Col. Noria Alta, Guanajuato, Gto., México, 36050

Research and Professional Experience: Research focus on the development and implementation of systematic methodologies in the construction of generic mathematical models for the design, synthesis and understanding of chemical and biochemical products and processes. 37

scientific refereed and indexed publications, 5 book chapters and 1 book. Participation in several national and international scientific conferences. Reviewer for different indexed journals. Currently, Coordinator of the Postgraduate Program in Chemical Engineering at the Universidad de Guanajuato.

Professional Appointments: Member of the of the Mexican Academy of Research and Teaching in Chemical Engineering (AMIDIQ). Registered as accredited evaluator for the National Council of Science and Technology (Mexico). Member of the Thematic Network on Bioenergy (Mexico).

Honors: Recognized as National Researcher by the National System of Researchers (SNI, México). 2019 Certificate of Reviewer Recognition by Elsevier.

Publications from the Last 3 Years:

- Gasca-González, R., Prado-Rubio, O. A., Gómez-Castro, F. I., Fontalvo-Alzate, J., Pérez-Cisneros, E. S., Morales-Rodriguez, R., 2019. "Techno-economic analysis of alternative reactive purification technologies in the lactic acid production process." *Computer-Aided Chemical Engineering.* 46, 457-462. (doi: 10.1016/B978-0-12-818634-3.50077-1).
- García-García, J. C., Ponce-Rocha, J. D., Marmolejo-Correa, D., Morales-Rodriguez, R., 2019. "Exergy analysis for energy integration in a bioethanol production process to determine heat exchanger networks feasibility," *Computer-Aided Chemical Engineering.* 46, 475-480. (https://doi.org/10.1016/B978-0-12-818634-3.50080-1).
- Contreras-Vargas, C. A., Gómez-Castro, F. I., Sánchez-Ramírez, E., Segovia-Hernández, J. G., Morales-Rodríguez, R., Gamiño-Arroyo, Z., 2019. "Alternatives for the Purification of the Blend Butanol/Ethanol from an Acetone/Butanol/Ethanol Fermentation

- Effluent," *Chemical Engineering Technology*, 42, 5, 1088–1100. (https://doi.org/10.1002/ceat.201800641).
- Meléndez-Hernández, P. A., Hernández-Beltrán, J. U., Hernández-Guzmán, A., Morales-Rodríguez, R., Torres-Guzmán, J. C., Hernández-Escoto, H., 2019. "Comparative of alkaline hydrogen peroxide pretreatment using NaOH and $Ca(OH)_2$ and their effects on enzymatic hydrolysis and fermentation steps." *Biomass Conversion and Biorefinery*. In press. (https://doi.org/10.1007/s13399-019-00574-3).
- Arce-Alejandro, R., Villegas-Alcaraz, J. F., Gómez-Castro, F. I., Juárez-Trujillo, L., Sánchez-Ramírez, E., Carrera-Rodríguez, M., Morales-Rodríguez, R., 2018. "Performance of a gasoline engine powered by a mixture of ethanol and n-butanol," *Clean Technologies and Environmental Policy*. 20, 1929-1937 (https://doi.org/10.1007/s10098-018-1584-5).
- Ponce-Rocha, J. D., Sánchez-Ramírez, E., Segovia-Hernández, J. G., Gómez-Castro, F. I., Morales-Rodriguez, R., 2018. "Optimized sustainable molecular and purification process design framework: acetone-butanol-ethanol case study," *Computer-Aided Chemical Engineering*, 44, 385-390.
- Méndez-Alva, J. A., Perez-Cisneros, E. S., Rodriguez-Gomez, D., Prado-Rubio, O. A., Ruiz-Camacho, B., Morales-Rodriguez, R., 2018. "Computer-aided process simulation, design and analysis: lactic acid production from lignocellulosic residues." *Computer-Aided Chemical Engineering*. 44, 463-468. (https://doi.org/10.1016/B978-0-444-64241-7.50072-0).
- Prado-Rubio, O. A., Rodriguez-Gomez, D., Morales-Rodriguez, R., 2018. "Model-Based Approach to Enhance Configurations for 2G Butanol Production through ABE Process," *Recent Innovations in Chemical Engineering*. 11, 99-111. (https://doi.org/10.2174/2405520411666180501112354).
- Sánchez-Ramírez, E., Ponce-Rocha, J. D., Segovia-Hernández, J. G., Gómez-Castro, F. I., Morales-Rodriguez, R., 2018. "A framework for optimised sustainable solvent mixture and separation

process design," *Computer-Aided Chemical Engineering.* 43, 755-760. (https://doi.org/10.1016/B978-0-444-64235-6.50133-9).

- Pino, M. S., Rodríguez-Jasso, R. M., Michelin, M., Flores-Gallegos, A. C., Morales-Rodriguez, R., Teixeira, J. A., Ruiz, H. A., 2018. "Bioreactor design for enzymatic hydrolysis of biomass under the biorefinery concept," *Chemical Engineering Journal.* 347, 119-136. https://doi.org/10.1016/j.cej.2018.04.057.

- Álvarez del Castillo-Romo, A., Morales-Rodriguez, R., Román-Martínez, A. (2018). "Multiobjective optimization for the socio-eco-efficient conversion of lignocellulosic biomass to biofuels and bioproducts," *Clean Technologies and Environmental Policies.* 20, 603-620.

Juan José Quiroz Ramírez

Affiliation: CONACyT – CIATEC A.C. Centro de Innovación Aplicada en Tecnologías Competitivas

Education: ScD in Chemical Engineering

Business Address: Omega 201, Col. Industrial Delta, 37545 León, Gto. México

Research and Professional Experience: Researcher in the area of process design, analysis and optimization, recognized as National Researcher by the National System of Researchers (SNI, México). 17 scientific papers published in indexed journals, 2 book chapters. Participation in several national and international scientific conferences. Reviewer for different indexed journals.

Professional Appointments: Member of the American Chemical Society. Registered as accredited evaluator for the National Council of

Science and Technology (Mexico). Member of the Thematic Network on Bioenergy (Mexico).

Publications from the Last 3 Years:

- Sánchez-Ramírez, E., Quiroz-Ramírez, J. J., Hernández, S., Hernández, J. G. S., Contreras-Zarazúa, G., & Ramírez-Márquez, C. (2019). Synthesis, design and optimization of alternatives to purify 2, 3-Butanediol considering economic, environmental and safety issues. *Sustainable Production and Consumption,* 17, 282-295.
- Sánchez-Ramírez, E., Quiroz-Ramírez, J. J., & Segovia-Hernandez, J. G. (2019). Synthesis, Design and Optimization of Schemes to Produce 2, 3-Butanediol Considering Economic, Environmental and Safety issues. In *Computer Aided Chemical Engineering* (Vol. 46, pp. 157-162). Elsevier.
- Alcocer-García, H., Segovia-Hernández, J. G., Prado-Rubio, O. A., Sánchez-Ramírez, E., & Quiroz-Ramírez, J. J. (2019). Multi-objective optimization of intensified processes for the purification of levulinic acid involving economic and environmental objectives. *Chemical Engineering and Processing-Process Intensification,* 136, 123-137.
- Quiroz-Ramírez, J. J., Sánchez-Ramírez, E., & Segovia-Hernández, J. G. (2018). Energy, exergy and techno-economic analysis for biobutanol production: a multi-objective optimization approach based on economic and environmental criteria. *Clean Technologies and Environmental Policy,* 20(7), 1663-1684.
- Quiroz-Ramírez, J. J., Sánchez-Ramírez, E., Hernández-Castro, S., Segovia-Hernández, J. G., & Ponce-Ortega, J. M. (2017). Optimal planning of feedstock for butanol production considering economic and environmental aspects. *ACS Sustainable Chemistry & Engineering,* 5(5), 4018-4030.

Juan Gabriel Segovia-Hernández

Affiliation: Departamento de Ingeniería Química, División de Ciencias Naturales y Exactas, Campus Guanajuato, Universidad de Guanajuato (México).

Education: Doctor in Chemical Engineering

Business Address: Noria Alta S/N Col. Noria Alta, Guanajuato, Gto., México, 36050

Strong expertise in synthesis, design and optimization of (bio) processes. He has contributed to defining systematic methodologies to found, in a complete way, optimum sustainable and green processes for the production of several commodities. He also applied his methodologies to the production of biofuels and Bio-Based Building Blocks. Products of his research are more than 120 papers published in high impact factor indexed journals, 3 books with prestigious international publishers and three patent registers. In addition, he acts as a reviewer for over 25 top journals in chemical engineering, energy, and applied chemistry. For the pioneering work and remarkable achievements in his area of scientific research, he was National President of Mexican Academy of Chemical Engineering (2013-2015). Also, he is "Associate Editor" of "Chemical Engineering and Processing: Process Intensification Journal" (Elsevier), since 2019.

Representative Publications from the Last 3 Years:

- Segovia-Hernández, J. G., Sánchez-Ramírez, E., Alcocer-García, H., Quiroz-Ramírez, J. J., Udugama, I. A., Mansouri, S. S., 2020, Analysis of Intensified Sustainable Schemes for Biobutanol Purification, *Chemical Engineering and Processing: Process Intensification,* 147, 107737.
- Alcocer-García, H., Segovia-Hernández, J. G., Prado-Rubio, O. A., Sánchez-Ramírez, E., Quiroz-Ramírez, J. J., 2020, *Multi-Objective*

- *Optimization of Intensified Processes for the Purification of Levulinic Acid Involving Economic and Environmental Objectives. Part II: A Comparative Study of Dynamic Optimization Approaches, Chemical Engineering and Processing: Process Intensification*, 147, 107745.
- Sánchez-Ramírez, E., Ramírez-Márquez, C., Quiroz-Ramírez, J. J., Angelina-Martínez, A. Y., Vicente-Cortazar, V., Segovia-Hernández, J. G., 2020, *Design of Dividing Wall Columns Involving Sustainable Indexes for a Class of Quaternary Mixtures, Chemical Engineering and Processing: Process Intensification*, 148, 107833.
- Gómez-Castro, F. I., Segovia-Hernández, J. G., 2019, "*Process Intensification: Design Methodologies,*" de Gruyter (ISBN 978-3110596076).
- Contreras-Zarazúa, G., Sánchez-Ramírez, E., Vázquez-Castillo, J. A., Ponce-Ortega, J. M., Errico, M., Kiss, A. A., Segovia-Hernández, J. G., 2019, Inherently Safer Design and Optimization of Intensified Separation Processes for Furfural Production, *Ind. Eng. Chem. Res.*, 58, 6105 - 6120.
- Ramírez-Márquez, C., Contreras-Zarazúa, G., Martín, M., Segovia-Hernández, J. G., 2019, *Safety, Economic and Environmental Optimization Applied to Three Processes for the Production of Solar Grade Silicon, ACS Sustainable Chemistry & Engineering*, 7, 5355 - 5366.
- Vázquez-Castillo, J. A., Contreras-Zarazúa, G., Segovia-Hernández, J. G., Kiss, A. A., 2019, Optimally Designed Reactive Distillation Processes for Eco-Efficient Production of Ethyl Levulinate, *Journal of Chemical Technology & Biotechnology*, 94, 2131 – 2140.
- Ramírez-Márquez, C., Vidal-Otero, M., Vázquez-Castillo, J. A., Martín, M., Segovia-Hernández, J. G., 2018, Process Design and Intensification for the Production of Solar Grade Silicon, *Journal of Cleaner Production*, 170. 1579 – 1593.
- Sánchez-Ramírez, E., Ramírez-Márquez, C., Quiroz-Ramírez, J. J., Contreras-Zarazúa, G., Segovia-Hernández, J. G., Cervantes-

Jauregui, J. A., 2018, Reactive Distillation Column Design for Tetraethoxysilane (TEOS) Production: Economic and Environmental Aspects, *Ind. Eng. Chem. Res.,* 57, 5024 - 5034.
- Torres-Ortega, C. E., Ramírez-Márquez, C., Sánchez-Ramírez, E., Quiroz-Ramírez, J. J., Segovia-Hernández, J. G., Rong, B. G., 2018, *Effects of Intensification on Process Features and Control Properties of Lignocellulosic Bioethanol Separation and Dehydration Systems,* Chemical Engineering and Processing: Process Intensification, 128, 188 - 198.

In: Properties and Uses of Butanol
Editor: Arnaud M. Artois

ISBN: 978-1-53618-448-8
© 2020 Nova Science Publishers, Inc.

Chapter 3

BUTANOL OXIDATION REACTION: FROM PT SINGLE CRYSTAL TO DIRECT BUTANOL FUEL CELL

Guilhermina F. Teixeira[1], Tarso L. Bastos[1],
Enrique Herrero[2], Juan M. Feliu[2] and Flavio Colmati[1],*

[1]Instituto de Química, Universidade Federal de Goiás, Goiânia, Brazil
[2]Instituto Universitario de Electroquímica, Universidad de Alicante, Alicante, España

ABSTRACT

Most part of the global energy is coming from fossil resources. Besides not being a renewable energy source, the products of their combustion processes are harmful to the environment. Fuel cells are a great renewable alternative to be used in energy production. The electricity generated by fuel cells is the result of the direct energy conversion promoted by the electrochemical transformation of compounds that store large amounts of hydrogen. Alcohols are often used as hydrogen sources

---

* Corresponding Author's Email: colmati@ufg.br.

and the chemical reactions that lead to the power generation can be catalyzed by precious metals. Platinum-based catalysts are efficient materials that can be used in the fuel cell operation. To synthetize the Pt catalysts, the most usual procedure consists of the reduction of soluble Pt species to produce nanoparticles with different morphology. The catalyst shape control is an important factor to consider since it results in atomic arrangements on the surface of the particle that leads to different exposed Pt active sites. These characteristics heavily depend on the synthetic method of the nanoparticles. After an adequate surface characterization, it is possible to establish a reliable mechanism for the oxidation reaction of hydrogen-rich compounds to generate energy and to find catalysts with higher activity for this reaction. Based on these contexts, in this chapter, we will discuss the oxidation of the butanol on Pt single crystal, the possible mechanism of the butanol oxidation reaction, and the working principles of fuel cells as well as the advantages of employing the Pt-based nanoparticles as catalysts in the energy generation process.

Keywords: direct alcohol fuel cell, butanol electrooxidation, single crystal, platinum, DAFC

INTRODUCTION

One of the main characteristics of our modern society is the extensive use of energy. The energy demands have increased exponentially from the beginning of the industrial revolution and the new developments are limited by the availability of energy. The oldest energy source known by humanity is the sun, which is still the major energy source in the world. The beginning of the human civilization is marked by the use of an additional energy source, the fire, for warming, preparing food and illumination. In that time, the fire was mainly obtained by burning wood from forests, which resulted in a negative environmental impact due to the devastation of forests in several places around the world. Due to the shortage of wood and the increasing human population, wood started to be replaced by coal for those purposes. Nevertheless, coal, like any other fossil resource, is a no renewable energy source. The advent of the Industrial Revolution lead to a continuous increase in the use of coal, initially, and later other fossil fuels to sustain the development of our society, which resulted in problems linked to the depletion of these resources as well as the environmental pollution caused by combustion process [1].

Nowadays the scientific and technological development has allowed us to build alternative devices to the non-renewable energy sources. So far, the sun is the biggest source of renewable and environmentally friendly energy. This fact has triggered the research for transforming solar energy into electricity in photovoltaic devices (solar cells) so that their light-harvesting capability is greatly increased [2].

Piezoelectric devices are another alternative in the production of renewable energy. Their operation is based on the piezoelectric effect that involves the conversion of mechanical energy into electricity (direct piezoelectric effect). The converse effect is responsible to convert electricity into mechanical energy. This behavior is related to the crystalline structure presented by the materials. The main characteristic to consider for the application of a given material in piezoelectric devices is to be constituted by a non-centrosymmetric structure. The low symmetry favors the dielectric displacement under the influence of mechanical stress which causes an internal polarization [3, 4]. The human body can be an abundant source of the mechanical stimulus [5], as well as wind [6] or waves marine [7].

An additional promissory way to produce clean energy is the direct conversion of chemical energy into electricity. The fuel cells convert the chemical energy from fuels in electrical work. The hydrogen is the lightest and one of the more abundant known elements in the world and due to its chemical properties, it is very adequate to produce energy. The most effective mean to produce energy from hydrogen is through the conversion of its chemical energy into electricity in fuel cells. The electricity generated by the fuel cells is the result of the electrochemical conversion of two coupled reactions: the oxidation reaction of a fuel, such as hydrogen or low molecular weight alcohols, and the oxygen reduction reaction. In this latter reaction, oxygen is taken normally from the air. These devices work since the global electrochemical reaction is a thermodynamically spontaneous process [8].

The fuel cell is generally composed of two metallic nanoparticles deposited on appropriate conductive supports which act as electrodes (cathode and anode) and the electrolyte. The oxidation of the fuel takes place at anode generating electrons that migrate by an external circuit while producing an electrical work to the cathode, where are employed in reducing an oxidizing agent. It is important to avoid the direct mixing of the reactants or the use of wrong electrodes because in both cases, there is a significant diminution of the production of electrical energy due to the formation of thermal energy. A solution for excluding accidental contact between electrodes is to insert an electronically insulating porous separator between them. To maintain the fuel cell working, the reactants should be continuously supplied to each electrode, and the formed products and residual heat should be removed [8].

The compounds used as fuels in the cells are reducing agents. There are several promising organic and inorganic compounds that are adequate candidates to be used as fuels [8]. The fuel cell using hydrogen at anode and oxygen at cathode produces the highest power density, but these gases must be of high purity to avoid deactivation of electrodes [9, 11]. As oxidizing agents, air or pure oxygen are commonly used, but other oxidizing agents such as hydrogen peroxide [8, 12] hydrazine [8, 13] and chlorine [8, 14] can also be employed. In the typical example of a fuel cell, the hydrogen-oxygen fuel cell in an acidic electrolyte, the following reactions are taking place:

Anode: $2H_2 \rightarrow 4H^+ + 4e^-$ (1)

Cathode: $O_2 + 4H^+ + 4e^- \rightarrow 2H_2O$ (2)

The formed protons in the electrolyte layer next to anode diffuse to the cathode through electrolyte resulting in a closed electrical circuit [8].

The major problem of using hydrogen is that it is not a readily available fuel in nature. Hydrogen may be considering an energy vector, which can be generated from several sources and transported where needed. However, the scarce hydrogen distribution infrastructure is a problem for the fuel cell commercialization [15]. Hydrocarbons and low weight molecular alcohols can replace hydrogen in fuel cell technology because they can be obtained from renewable sources and can be distributed using the present infrastructure. Their electrochemical oxidation reaction for energy generation is catalyzed by precious metals at the anodes. Different metals and metal alloys are used as catalysts. As in any other heterogeneous catalyzed reaction, the reaction rate depends on the interactions between the reactant or some reaction intermediates with the surface of the catalyst, so that the activation energy for the desired reaction is lowered. Thus, a direct interaction between the atoms in the surface of the catalysts and the species participating in the reaction is required [16]. In this interaction with the surface, not only the composition of the catalyst, that is the nature of the atoms on the surface, but also the geometrical distribution of the surface atoms (the surface structure) affects the catalysis. The atoms located on the surface of the catalyst are unsaturated when compared to those in the bulk of the metal so that they are available for the adsorption of molecules present in the vicinity of the surface [17]. When the adsorption energy and geometry are the correct ones, the adsorption process leads to the activation of these molecules. In a fuel cell, the fuel adsorbs on the metal surface of the catalyst, where it is activated, and reacts on the surface to produce adsorbed products that are subsequently desorbed from the metal surface to yield the desired products, as showed in the scheme represented by Figure 1. In this figure the blues arrows represent the interaction between the atoms in bulk and red arrows indicate the surface sites that can interact with the adsorbate.

These catalytic events depend heavily on the different crystallographic organization on the exposed surface of a material. Hence different crystallographic orientation, composition, and morphologic variety result in different catalytic activity for a reaction [18]. Different surface occupancy results in different coordination numbers and consequently offer different electronic states on the surface, which give rise to different adsorption energies of a compound on the surface of the metallic particles during the catalytic process. This adsorption energy can be modified by the surface occupancy of the adsorbed molecule because of the presence of attractive or repulsive interactions between adjacent adsorbates [19]. All of these factors affect the adsorption energy or activation barrier for the dissociation of a reactant molecule. In metals with a face-centered cubic crystallographic structure, the surface occupancy of surface metal follows the order: (111) >

(100) > (110). However, when a metal catalyst has a body-centered cubic lattice the order is (100) > (100) > (111). To understand how the surface structure affects the electrochemical activity, studies with single-crystal electrodes are normally carried out. The initial studies are normally conducted using the low index planes because they contain only one type of site. However, only a full picture of the effect of the surface structure in the catalysis is obtained when stepped and kinked surfaces are used because these surfaces combine, in a controlled disposition, two or three types of sites [20, 21].

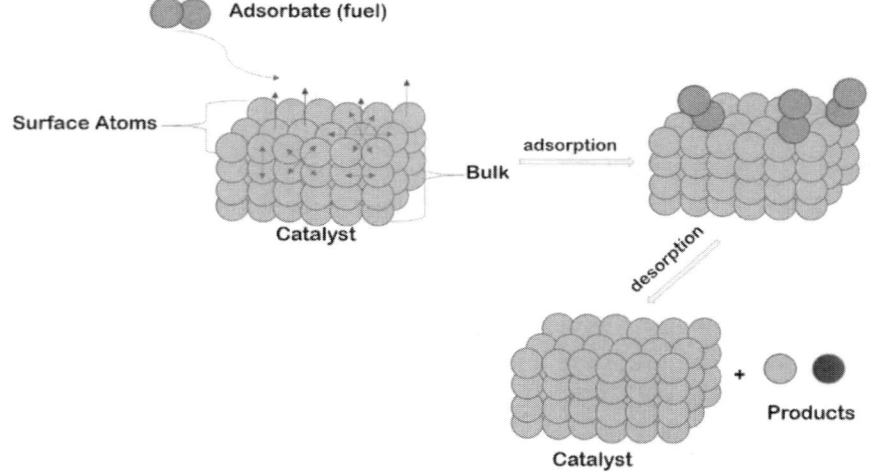

Figure 1. Schematic illustration the adsorption-desorption processes on catalyst surface in a general fuel cell.

Using this approach, a detailed analysis of the effect of the surface structure can be achieved. Since single crystal surfaces cannot be used in practical applications, this analysis has to be extended to the practical electrodes, those composed by nanoparticles. The atomic arrangements and surface structure are determined by the crystalline and morphologic structure of the catalyst. The density and nature of the active sites are linked to the shape of the catalyst. Thus, when particles with the same volume and different morphology are compared to each other, spherical-shaped particles have the smallest surface area resulting in the lowest percentage of surface atoms [17].

Due to the shape-dependent crystallographic surface, there has been an increasing interest in the synthesis of metallic catalysts with a specific morphology. The control of the morphology of metal nanoparticles can be used to explore how surface structure can affect the selectivity and catalytic performance of metallic nanoparticles so that the reactivity for the desired reaction is maximized [22]. Moreover, the use of nanoparticles with a controlled shape is mandatory when the reaction mechanism in nanoparticles is to be untangled because correlations between the surface structure and shape of the nanoparticles and the reactivity for the desired reaction can be obtained [23]. The different crystallographic planes have different energy levels, coordination numbers, and distances between adjacent atoms resulting in different adsorption energy for the studied molecule.

Nanoparticles with a surface composed by higher Miller index have demonstrated high activity for many reactions; however, their synthesis and their long-term stability are still a challenge [19]. In the study conducted by Ferreira et al. (2007) faceted and dendritic Pt nanoparticles were produced by reducing Pt soluble species in proton exchange membrane fuel cells. The morphology of Pt single-crystal nanoparticles evolves from dendritic to truncated tetrahedrons, truncated octahedrons, and truncated square cuboids. The shape of Pt nanoparticles was determined by a competition between interfacial kinetics and surface energy of Pt reduction following by Pt incorporation onto different surfaces. For a metal like Pt that presents a face-centered cubic structure, the surface energy of atomic planes with high symmetry follows the order $\{111\}_{Pt} < \{100\}_{Pt} < \{110\}_{Pt}$. In addition, based on calculated surface energies, the authors present a Wulff geometrical construction showing the equilibrium crystal shape [24]. Catalytic reactions can be improved using catalysts with high index planes because of their better catalyst activity and selectivity. Pt concave nanocubes with high-index facets were obtained by Yu et al. (2011). The surface of nanocubes were mainly composed by $\{720\}$ facets together $\{510\}$, $\{830\}$ and $\{310\}$ facets. The processing route was the key role to form Pt concave nanocubes and block the growth of the <100> axis favor the growth of facets high

Miller index [25]. The review Cao et al. clearly illustrate the influence of size, shape and nanocatalyst surface on the oxidation and reduction reactions, both on an organized surface with low-Miller indices and on that composed by high-Miller index [19].

Pt-based catalysts are the most studied for fuel cell applications their good adsorption properties for the studied fuels. Pt can be used in a single, double, or multiple-component catalyst [26-30]. The application of binary and ternary Pt alloys in electrochemical reactions can improve the catalytic performance due to the synergetic effect promoted by the metal combinations [31]. The electrocatalytic activity of Pt-based catalysts is significantly linked to the chemisorption properties of the surface atoms so that there is a decrease of the free-energy barriers for the reaction [31, 32]. In electrochemical reactions, like those taking place in fuel cells, the electrode surface composition is an important parameter to be considered. As aforementioned the catalytic activity on Pt alloys depends on their crystalline orientation. The surface of conventional Pt alloys consists of different crystalline structures and defects that may result in low activity. The control of the shape of Pt alloys allows maximizing the reaction rate. The reviews by Shao et al. (2016) [33] and Li et al. (2018) [31], provide a detailed report on the surface characteristics of the Pt alloys depending on synthetic method and its performance for some reactions. As an example, a bimetallic catalyst based on Pt nanoparticles has improved catalytic activity in the oxygen reduction reaction. This behavior is attributed to the electronic events promoted by the surface atomic arrangement in the exposed facets of special nanoparticles. Li et al. (2015) investigated the growth mechanism of bimetallic Pt_xCu_y nanocrystals with hexapod morphology. The hexapods have exposed (111) and other facets with a high index and present a high ratio of atoms at the corner and edges. Due to these characteristics, the catalytic activity towards the oxygen reduction reaction was notoriously improved using Pt_xCu_y bimetallic nanocrystals [34].

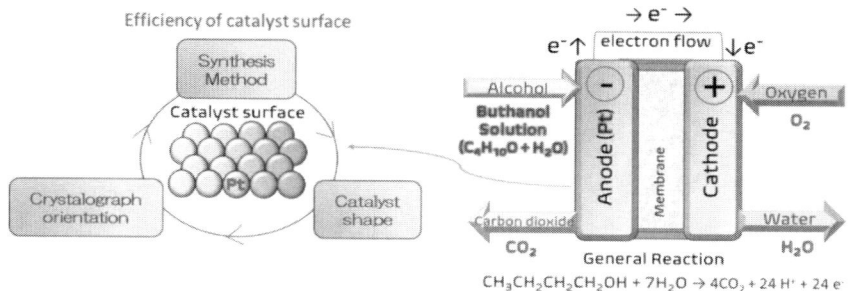

Figure 2. Schematic illustration of a Pt-based fuel cell supplied by butanol as fuel.

Fuel cells have been proven to be a great technology to produce clean energy. In contrast to solar cells and energy harvesting based on the piezoelectric effect, the fuel cells do not stock energy, but they deliver in the form of electricity straight from a fuel. Furthermore, fuel cells should be supplied with oxygen and fuel to operate. Several kinds of fuel cells have been applied in energy production, such as proton exchange fuel cells [35], molten carbonate fuel cell [36], phosphoric acid fuel cell [37], alkaline fuel cell [38] and direct alcohol fuel cell (DAFCs) [39]. The DAFCs works using alcohol as fuel and metal are used as electrocatalysts. The studies on the DAFCs based on methanol or ethanol are very numerous, however, both alcohols present some disadvantages in their use. With the goal of obtaining a more efficient cell and minimizing the disadvantages of the use of methanol and ethanol alone, Leo et al. (2011) built a fuel cell based on a mix of methanol and ethanol in different ratios, and they conclude that a fuel cell with a mixed fuel is a field that is worth exploring [40]. Another option to replace the methanol and ethanol in fuel cells is the use of other alcohols, such as butanol, as combustible. Even though butanol can be a very good option for this purpose its use is still little explored. As explained before, the efficiency of fuel cell is related to catalyst properties, such as shape of the metallic particles and crystallography orientation of the catalyst surface. In turns, these factors are controlled by employed synthetic method to produce the catalyst material. In Pt-based fuel cell supplied with butanol, the alcohol is adsorbed on Pt surface, where the oxidizing reaction takes place. These concepts may be represented by the scheme showed in Figure 2. So, the present chapter covers the research about the direct fuel cells based on butanol and the use of Pt as a catalyst in its electrochemistry reaction.

OXIDATION OF BUTANOL ON SINGLE CRYSTALS AND METAL ALLOYS

The studies of Direct Alcohol Fuel Cells (DAFCs) for electricity generation are very abundant because this technology uses renewable liquid fuels like methanol, ethanol, etc., which present less difficult for handling, storage, transport, and have high energy density than hydrogen and can be used as fuel in fuel cells [41, 42]. However, the utilization of some of these molecules in fuel cells has some disadvantages. Methanol is toxic and has a low boiling point (65°C) [43]. Ethanol is a good alternative because of its physical-chemical characteristics, but, on the other hand, it is obtained from food stocks, which may compromise the food supply for humanity, and the complete ethanol oxidation reaction has not been achieved because the cleavage of the C-C bond is a difficult step. [44, 45 46] Other alcohols have also been considered for fuel cells, but they have deserved much less attention. Butanol is one of those. It is a long-chain alcohol, which has a high energy density and can be produced from biomass [41, 42]. It presents several isomers, whose oxidation behavior has been studied on various electrodes [47], to determine the possible mechanism of oxidation.

Long-chain alcohols are more difficult to oxidize to CO_2 because of numerous intermediates and adsorbate species that are formed, and thus reducing the efficiency of the fuel cell [48, 49]. Thus, the number of electrons obtained for the oxidation of the alcohol is lower than the theoretical electron yield for a complete conversion to CO_2. As aforementioned, the studies carried out with ethanol demonstrate that the most difficult step is the cleavage of the C-C bond and so that acetaldehyde and acetic acid are formed instead of CO_2. The formation of acetaldehyde and acetic acid involves the exchange of 2 e^- and 4 e^-, respectively, in contrast to the 12 e^- when CO_2 is formed as the final product [42, 50]. Li et al. show that the reactivity of primary alcohols has the following order: methanol > ethanol > propanol > n-butanol in acid media on a Pt electrode [51]. Furthermore, during the alcohol oxidation, CO, which is strongly adsorbed and poisons the Pt catalysts, can be formed. Also, C_2, C_3, and C_4 intermediates may be strongly adsorbed on the surface diminishing its reactivity to the desired reaction [52].

Takky et al. (1983) studied different noble metallic polycrystalline electrodes for the oxidation of the isomers of butanol in alkaline medium using cyclic voltammetry. In alkaline medium, n-butanol, isobutanol, s-butanol, and t-butanol were studied on Pt, Au, Pd, and Rh [53]. On platinum electrodes, n-butanol and isobutanol have very similar behavior, with peak current densities respectively of 0.4 and 0.45 mA cm^{-2}, and the main oxidation peak centered at ca. 0.83 V (vs. RHE). The s-butanol oxidation process shows four different peaks at 0.35, 0.49, 0.73 and 1.1 V, and peak current densities of 0.35 mA cm^{-2}. On gold, the n-butanol, isobutanol, s-butanol has a similar profile with the same peak at −1.23 V and 1.08 V during the positive and negative scan directions, respectively. On the other hand, s-butanol has the highest peak current densities when compared to those of n-butanol and isobutanol, (6.1 mA cm^{-2} vs. 3.4 mA cm^{-2} and 2.7 mA cm^{-2} respectively). Isobutanol was the most reactive isomer on Pd electrodes, with a peak current of 0.8 mA cm^{-2} at 0.8 V and 1.6 mA cm^{-2} at 0.71 V in the negative and positive scan directions, respectively. The reactivity of n-butanol on Pd (0.5 mA cm^{-2} at 0.85 V/MSE) is similar to that on Pt, while that of s-butanol is much lower (0.15 mA cm^{-2} at 0.73 V). Rhodium is very unactive for the electro-oxidation of butanol n-butanol, isobutanol, and t-butanol. Only for s-butanol, a small oxidation peak (0.18 mA cm^{-2} at 0.25 V) is observed. The t-butanol is not oxidized on Au and Pd during positive scan direction, and on Pt and Rh on both directions [53]. This study highlights that the dependence on the molecular structure of electrochemical oxidation of butanol in an alkaline medium can be caused by steric effects, inductive effects, or both [54]. In alkaline medium, gold displays the highest currents but at a very high potential, which precludes its use in fuel cells. The highest current density on gold electrode oxidation is for the s-butanol, and thus, the secondary alcohol functional group is more easily oxidized than the primary alcohol functional group, while these differences are not observed on platinum [53].

This previous study shows that platinum proved some efficiency in the oxidation of n-butanol, isobutanol, and s-butanol, in both alkaline and acid medium at low potentials, which is the desired potential window for their use in fuel cells. Moreover, their oxidation depends on the molecular structure of the isomer [54]. Takky et al. (1985) propose that the inductive electronic effect explains the current toward the butanol oxidation reaction, where the bond breaking between the α-carbon and hydrogen atom is a limiting step. Thus, the t-butanol is not being oxidized at room temperature because it has no hydrogen on the α-carbon, and, being a tertiary carbon, it is the most stable of the four butanol isomers [54]. The above experiments were carried out employing polycrystalline metals electrodes, although the effect of the crystalline structure of electrode was demonstrated as an important effect in the oxidation of alcohol molecules [54] where the highest current density depended on the single crystal plane, and others differences were observed when changing the solvent and index crystal in ethanol oxidation [55]. Thus, it can be assumed that the molecular structure of isomers and crystalline structure of electrodes directly influence the efficiency of the redox reactions as an energy source.

In another study Takky et al. (1988) showed the influence of crystal structure of platinum for the oxidation of the butanol isomers in alkaline medium, using the low-index planes, i.e., Pt (100), Pt (110), and Pt (111), and comparing their behavior with that observed for polycrystalline platinum [56]. The electro-oxidation of n-butanol shows that the current densities recorded for Pt (100) (3.2 mA cm^{-2}) are higher than those measured in polycrystalline electrodes (2.8 mA cm^{-2}), and both were smaller than the ones obtained for the Pt (110) (12.2 mA cm^{-2}) and Pt (111) (18.3 mA cm^{-2}). The maximum current densities were obtained in the first cycle of voltammograms, but they were also compared with those in the fifth cycle to analyze the poisoning effect. For all of the electrodes, a significant diminution was observed upon cycling. The values obtained were: Pt (poly) 1.69 mA cm^{-2} (40% decrease in the fifth cycle), Pt (100) 1.22 mA cm^{-2} (58%), Pt (110) 3.6 mA cm^{-2} (70%) and Pt (111) 12.0 mA cm^{-2} (34%). These results show that Pt (111) is the most active orientation and also the one having the best resistance to the poisoning phenomena [56]. The isobutanol oxidation reaction behaves similarly to oxidation of n-butanol with a smaller current density for all the orientations, where the peak currents for the first cycle are: Pt (poly) 2.6 mA cm^{-2}, Pt (100) 2.8 mA cm^{-2}, Pt (110) 5.2 mA cm^{-2} and Pt (111) 11 mA cm^{-2}. They also undergo a significant poisoning effect with a decreasing current density after fifth cycle: Pt (poly) 1.22 mA cm^{-2} (53%), Pt (100) 1.48 mA cm^{-2} (47%), Pt (110) 2.8 mA cm^{-2} (46%) and Pt (111) 7.2 mA cm^{-2} (34%). s-butanol has a smaller current density than that observed for the primary isomers with Pt (poly) 0.93 mA cm^{-2}, Pt (100) 1.8 mA cm^{-2} and Pt (110) 1.6 mA cm^{-2}, although the current for Pt (111) is larger than that obtained for isobutanol (11.8 mA cm^{-2}). However, the main difference between s-butanol and the other isomers was the better resistance to the poisoning effect, with a smaller decrease at the fifth cycle: Pt (poly.) 0.93 mA cm^{-2} (1.1%), Pt (100) 1.48 mA cm^{-2} (25%), Pt (110) 1.4 mA cm^{-2} (12.5%) and Pt (111) 11.1 mA cm^{-2} (6%) [56].

All these results demonstrate that the surface structure of the electrode affects significantly to the electrochemical oxidation of the butanol isomers, not only in the direct current density but also in the poisoning effect. The Pt (111) single-crystal plane was the most active with the highest current density, and the highest resistance to the poisoning effect. Thus, the reaction order for the different electrodes according to their activity is Pt (111) >> Pt (110) > Pt (100) > Pt (Polycrystalline) [57]. Tripković et al. also study the Pt (111) electrode and the Pt (755) and Pt (332) vicinal single-crystal planes, and the obtained surface activity order was Pt (332) > Pt (755) > Pt (111) [57]. Nevertheless, the structure of the isomer also influences the oxidation during the continuous cycling, where the most reactive are the primary isomers, with the higher current density, but also undergoing a strong poisoning effect and a significant decrease in reactivity. Then the most viable option to use in devices is s-butanol with (111) facets (11.8 mA cm^{-2}) because it has a higher current density and a lower poisoning effect. If only the direct activity is considered, the obvious choice is an electrode composed of nanoparticles with (111) facets oxidizing n-butanol (18.3 mA cm^{-2}) [56]. However, for practical application in fuel cell devices, it is necessary to evaluate real currents and the poisoning effect, so that the higher current density simultaneously with a weak poisoning effect in prolonged operation is obtained.

The utilization of a single noble metal electrode can have disadvantages during a redox reaction, like the poisoning effect, the decrease of reactivity, and the adsorption of intermediates that block the desired reactions [58, 59]. On the other hand, the addition of a second or third metal in a Pt alloy, as Cu or Sn, can increase the long term poisoning resistance, and the activity when compared to the monometallic catalyst [42, 59]. It has been reported that the PtRu is the best bimetallic catalyst for methanol oxidation and PtSn for ethanol oxidation [41]. Thus, many authors have studied the oxidation of butanol in Pt alloys in order to obtain the best efficiency. Puthiyapura et al. started studying the influence of Sn in a bimetallic catalyst for n-butanol oxidation in acid medium and compared with Pt [41]. The voltammograms show the characteristic oxidation potential peak of n-butanol for both electrodes, however, the onset potential for PtSn ($\sim$ 0.03 V) is lower than that of Pt ($\sim$ 0.55 V). Besides that, the peak current density for n-butanol oxidation on pure Pt is ~2-3 times higher than that measured on PtSn. This effect can be attributed to the addition of Sn, where the alloy showed an increase of dissociative adsorption efficiency [41]. Additionally, Puthiyapura studied Pt and PtSn catalysts for the oxidation of butanol isomers in acidic medium. A lower peak potential for the oxidation of iso-butanol in PtSn was also observed, having a very similar behavior to n-butanol [41]. These enhancements in the oxidation are not observed for s-butanol, where Pt and PtSn showed very similar behavior. As before, the oxidation of t-butanol is completely inhibited in both metal catalysts. In conclusion, PtSn alloys have higher activity for the oxidation of n-butanol and iso-butanol at low potentials, while no enhancement is observed for the oxidation of s-butanol and [39]. These results indicate that butanol isomers have different oxidation mechanisms, where the catalyst can undergo a poisoning effect of different intensity, depending on the specie that is adsorbing during the oxidation process, and thus changing fuel cell efficiency.

BUTANOL ELECTRO-OXIDATION PATHWAY AND THE FORMED PRODUCTS

The widespread use of fuel cells as electricity source in devices depends on efficiency when using an alcohol fuel as a substitute to the hydrogen gas. As shown earlier many studies modify the catalyst and use different alcohols to obtain the best current density and the lowest poisoning effect. However, its efficiency depends on the mechanism and the formed product during the oxidation process, while the partial oxidation and adsorbed species decrease the efficiency and lifespan of the fuel cell devices.

In a study on the development of DAFCs and their reaction mechanism, Lamy et al. used thermodynamic data for the oxidation processes to calculate the thermodynamic cell potential under standard conditions. The values of E_{cell}° were 1.213 V, 1.145 V, 1.067 V, and 1.029 V, and a reversible energy efficiency at equilibrium potential of 0.967, 0.969, 0.916, and 0.890 for methanol, ethanol, propanol, and butanol respectively [50]. These numbers show that the increase in the carbon chain decreases the standard potential and the energy efficiency of the alcohol oxidation. However, the thermodynamic values do not take into account kinetic rates and side reactions that may lead to poisoning effects. As previously discussed, the molecular structure directly influences the measured current density and the poisoning effect, and in some cases, even prevents the measurement of an actual oxidation current. All these factors are related to the oxidation reaction mechanism of the different alcohol molecules. The determination of a full molecular mechanism is a very complicated task because it requires the identification, for each electrode, of the intermediates involved in the reaction. In the identification of the intermediate species, infrared spectroscopy, UV-VIS and "on line" chromatographic has been widely used [50].

The complete alcohol oxidation to CO_2 always requires the transfer of oxygen groups to the molecule. This oxygen group can come from a water molecule or from adsorbed OH on the electrode surface, as shown by the general equation for the electro-oxidation of a primary alcohol (3).

Therefore, the catalyst must be able to activate the water molecules of the solution and the adsorbed species in a complex mechanism, in which several adsorbed intermediates and products and by-products are formed [50].

$$C_nH_{2n+1}OH + (2n-1)H_2O \rightarrow nCO_2 + 6nH^+ + 6ne^- \qquad (3)$$

Equation 3 indicates that the complete oxidation of methanol involves 6 e^-. However, carbon monoxide is formed as an intermediate in the reaction and acts as the main poisoning species because it is strongly adsorbed on platinum electrodes, blocking more than 90% of the active sites. Additionally, other adsorbed species, such as $(\cdot CHO)_{ads}$ and $(\cdot COOH)_{ads}$ have been proposed [48]. The methanol oxidation mechanism, which is a six electron reaction, is already complex and involves the formation of formaldehyde and formic acid as intermediates to form CO_2 as the final product. The study of intermediates is crucial to understand the mechanism and to prevent the formation of adsorbed species that block the active sites of catalyst, mainly CO, decreasing the fuel cell efficiency [50].

As the number of carbon increases, the number of exchanged electrons also increases, and the elucidation of the mechanism becomes more complicated. The complete oxidation reactions of ethanol, propanol, and butanol involve 12 e^-, 18 e^- and 24 e^-, respectively. The ethanol oxidation mechanism has been intensively investigated among these alcohols [60, 61], on pure platinum and several alloy electrodes (Pt/X with X=Ru, Sn, Mo) [50]. Intermediates, mainly CO, are adsorbed rapidly in pure platinum electrodes as the result of the chemical dissociation of the molecule [50]. The ethanol oxidation process can generate different products, as acetic acid at high electrode potential (E > 0.6 V/RHE) according to reaction (4), or generate acetaldehyde, (reaction (5)), at lower potentials (E < 0.6 V/RHE) [50, 60, 61].

$$CH_3-CH_2OH + H_2O \rightarrow CH_3-COOH + 4H^+ + e^- \qquad (4)$$

$$CH_3-CH_2OH \rightarrow CH_3-CHO + 2H^+ + 2e^- \qquad (5)$$

On the active platinum sites, the activation of water to form adsorbed OH takes place (reaction (6)) ($0.6 \leq E \leq 0.8$ V/RHE), and thus, the oxygen group required for the oxidation of acetaldehyde to acetic acid is available on the Pt surface so that the oxidation can take place according to reactions (6) and (7) [50].

$$Pt + H_2O \rightarrow Pt-OH_{ads} + H^+ + e^- \qquad (6)$$

$$(CH_3-CHO) + Pt-OH_{ads} \rightarrow CH_3-COOH + H^+ + e^- + Pt \qquad (7)$$

The oxidation to form carbon dioxide on pure Pt is difficult at room temperature because it requires the cleavage of the C-C bond. However, this bond braking has been observed by gas chromatography and infrared reflectance spectroscopy at lower potential ($E < 0.4$ V/RHE). Then, a possible mechanism for the formation of carbon dioxide is the following [62, 63].

$$Pt + CH_3CHO \rightarrow Pt-(CO-CH_3)_{ads} + H^+ + e^- \qquad (8)$$

$$Pt + Pt-(CO-CH_3)_{ads} \rightarrow Pt-(CO)_{ads} + Pt-(CH_3)_{ads} \qquad (9)$$

$$2Pt + H_2O \rightarrow Pt-H_{ads} + Pt-OH_{ads} \qquad (10)$$

$$Pt-(CH_3)_{ads} + Pt-H_{ads} \rightarrow CH_4 + 2Pt \qquad (11)$$

$$Pt-(CO)_{ads} + Pt-OH_{ads} \rightarrow CO_2 + H^+ + e^- + 2Pt \qquad (12)$$

The presence of adsorbed species causes the inhibition of the reaction and the decrease of the efficiency and lifespan of the fuel cell. When other metals are used as electrocatalyst (rhodium, iridium, and gold), they show similar behavior in acid medium, forming carbon monoxide and acetic acid, but at higher potential values and with lower efficiency than pure platinum [50, 60, 61]. The most difficult steps for the rection are the oxidation of adsorbed CO at lower potential and the cleavage of the C-C bond at low temperatures. In IR spectroscopy, the adsorbed CO shows two bands, related to and bridge-bonded CO. Generally, the most intense band is the one related to linearly-bonded CO at ca. 2000 cm^{-1} [60, 61].

Li and Sun (1997) studied the n-butanol electro-oxidation on Pt electrodes in perchloric acid medium, using in situ Fourier-transform infrared spectroscopy (FTIR) to determine the oxidation mechanism [51, 64]. The FTIR spectra show the bands at 1104 cm^{-1} and 1600 cm^{-1}, associated with ClO_4^- adsorption on the surfaces and H_2O bending modes (1645 cm^{-1}), respectively. At low potentials, (between 0.0 V and 0.4 V) bands at ca, 1600 cm^{-1} assigned to C=C groups where observed, which indicates that the dehydration of n-butanol is taking place at those potentials according to [64].

$$CH_3CH_2CH_2CH_2OH \rightarrow CH_3CH_2CH=CH_2 + H_2O \tag{13}$$

For E < 0.3 V a small band close 2060 cm^{-1}, characteristic adsorption band of linear bonded CO was also observed [64]. Thus, the CO adsorption should occur during butanol oxidation, as follows:

$$CH_3CH_2CH_2CH_2OH + Pt \rightarrow Pt-CO_{ads} + CH_3CH_2CH_3 + 2H^+ + 2e^- \tag{14}$$

When the potential is increased from 0.3 V, two bands were observed in the spectra at 1712 cm^{-1} related to the carbonyl group (C=O) and at 2345 cm^{-1}, indicative of CO_2 formation. Simultaneously the bands related to adsorbed CO and the presence of C=C groups diminish, which suggests the oxidation of C=C species and CO to generate carbonyl groups and CO_2, respectively. The intensity of ClO_4^-, C=O, and CO_2 bands increased in $E_2 = 0.4$ V while the CO band decrease and a new band appears near 1003 cm^{-1}, attributed to out-of-plane OH$\cdots$O vibrational mode [62, 64].

At a high potential, $E_2 > 0.4$ V, the band associated with adsorbed CO disappears, which indicates the complete oxidation of CO in CO_2. Thus, CO is no longer adsorbed on the Pt surface, allowing the butanol oxidation reaction and other intermediate species can be identified by FTIR [64]. At those potentials, the evolution of the H$_2$O band at 1650 cm^{-1} indicates the consumption of water in the oxidation reaction of the different identified species [62, 64]. In the complete oxidation mechanism of all these species, CO$_2$ is formed as final species, according to

$$CH_3CH_2CH=CH_2 + 8H_2O \rightarrow 4CO_2 + 25H^+ + 25e^- \qquad (15)$$

$$Pt-CO_{ads} + H_2O \rightarrow Pt + CO_2 + 2H^+ + 2e^- \qquad (16)$$

$$CH_3CH_2CH_2CH_2OH + 7H_2O \rightarrow 4CO_2 + 24H^+ + 24e^- \qquad (17)$$

However, the small intensity of the bands at 1650 cm^{-1} and that related to the formation of CO$_2$ (2345 cm^{-1}) indicates that equations (15) and (17) are unlikely to occur [43]. In this respect, the two bipolar bands near 2960 cm^{-1} and 2885 cm^{-1} associated with the C−H stretching modes of CH$_3$ and CH$_2$, respectively also observed, as well as, and the bands at 1396 cm^{-1} assigned to C−OH bending, at 1273 cm^{-1} for OH deformation and 1211 cm^{-1} for the C−O stretching, all characteristics of the −COOH functional group. Based on observations, the most probable oxidation product of n-butanol is the corresponding acid species (R−COOH) [64].

Based on all results and discussions, it can be said that there are two paths for the complete oxidation mechanism of butanol to carbon dioxide. The first path goes through poisoning intermediates, mainly CO, that yields CO$_2$. In the second path through several intermediates, such as CH$_3$CH$_2$CH$_3$, CH$_3$CH$_2$CH=CH$_2$ and CH$_3$CH$_2$CH$_2$CHO are formed, and some of them are eventually oxidized to CH$_3$CH$_2$CH$_2$COOH and/or undergo the difficult and slow C−C breaking step to form one carbon species, which subsequently transforms into CO and CO$_2$ [64].

Finally, butanol could be presented as an alternative toward direct alcohol fuel cell to convert chemical energy into electrical energy moreover, it is a hydrogen-rich compound, although an adequate electrocatalyst should be developed for complete butanol oxidation with high electrochemical rate and low poisoning effect [42, 65]. Butanol can be a promising fuel once it may provide 24 electrons per molecule when the complete oxidation is achieved (as represented by Equation 17 and Figure 2).

CONCLUSION

The energy demands have increased exponentially from the beginner of the industrial revolution among several energy sources a promissory way to produce clean energy is fuel cells that convert directly chemical energy into electricity. The fuel cell is composed of metallic nanoparticles deposited on a conductive support and the electrolyte. When the fuel cell woks at low temperatures the electrodes are composed by precious metals, which deactivate be the adsorption of by-products. The different crystallographic orientation, composition, and morphologic variety of the metallic electrodes result in different catalytic activity for a reaction toward butanol oxidation reaction. These effects have been shown for the butanol oxidation reaction using Pt single crystal electrodes, which show different electrochemical activity toward butanol oxidation reaction. The reactivity order toward butanol oxidation on Pt single crystal electrodes is, according to the maximum current density: Pt (111) >> Pt (110) > Pt (100) > Pt (Polycrystalline) according to current density. On the other hand, the oxidation rate is also dependent on the isomer. The experimental results indicate that the reaction rate is proportional to the facility to extract the α-hydrogen from the molecule. Thus, currents for the oxidation of t-butanol are the lowest, because of the absence on a hydrogen atom in the α position.

REFERENCES

[1] Science Clarified. (2020). *The Development of Energy*. Accessed June 25. http://www.scienceclarified.com/scitech/Energy-Alternatives/The-Development-of-Energy.html.

[2] Chen, G., Zhijun, N., Hans Å. 2016. "Nanostructured Solar Cells." *Nanomaterials* 6:6-8.

[3] Tichý, J., Erhart, J., Kittinger, E., Privratská, J. 2010. "Principles of Piezoelectricity." In: *Fundamentals of Piezoelectric Sensorics: Mechanical, Dielectric, and Thermodynamical Properties of Piezoelectric Materials*, edited by Jan Tichy, Jiri Erhart, Erwin Kittinger and Jana Privratska, 1-14. Berlin: Springer - Verlag Berlin Heidelberg.

[4] Soin, N., Anand, S. C., Shah, T. H. 2016. "Energy Harvesting and Storage Textile." In: *Handbook of Technical Textiles. Volume 2: Technical Textile Applications*, edited by A. Richard Horrocks and Subhash C. Anand 357 - 396. Duxford, UK: Woodhead Publishing.

[5] Ahn, S., Cho, Y., Parka, S., Kima, J., Suna, J., Ahna, D., Leea, M., Kima, D., Kimc, T., Shinb, H., Parka, J.-J. 2020. "Wereable Multimode Sensors, with Amplified Piezoelectriciy due to the Multi Local Strain Using 3D Textile Structure for Detecting Human Body Signal." *Nano Energy* 2020:104932, doi.org/10.1016/j.nanoen.2020.104932.

[6] Tao, J. X., Viet, N. V., Carpinteri, A., Wang, Q. 2017. "Energy Harvesting from Wind by a Piezoelectric Harvester." *Engineering Structures* 133:74 – 80.

[7] Viet, N. V., Wu, N., Wang, Q. 201). "A Review on Energy Harvesting from Ocean Waves by piezolectric technology." *Journal of Modeling in Mechanics and Materials* 1:2016-0161.

[8] Bagatsky, V. S. 2012. "The Work Principle of a Fuel Cell." In: *Fuel Cells: Problems and Solution*, edited by Vladimir S. Bagatsky 5-24, New Jersey: A John Willey & Sons, Inc., Publications.

[9] Amamou, A. A., Kelouwani, S., Boulon, L., Agbossou, A. 2016. "A conmprehensive review of solutions and strategies for cold start of automotive proton exchange membrane fuel cells." *IEEE Access* 4 4989-5002.

[10] Zamel, N., Li, X. 2011. "Effect of contaminants on polymer electrolyte membrane fuel cells." *Progress Energy and combus sci* 37:292-329.

[11] Colmati, F., Lizcano-Valbuena, W. H., Camara, G. A., Ticianelli, E. A., Gonzalez, E. R. 2002. "Carbon monoxide oxidation on Pt-Ru

electrocatalysts supported on high surface area carbon." *Journal of the Brazilian Chemical Society* 13:474-482.
[12] Pan, Z., Bi, Y., Na, L. 2019. "Performance Characteristics of a Passive Direct Ethylene Glycol Fuel Cell With Hydrogen Peroxide as Oxidant." *Applied Energy* 250:846 – 854.
[13] Hwang, H., Hong, S., Kim, Do-H., Kang, Moon-S., Park, Jin-S., Uhm, S., Lee, J. 2020. "Optimistic Performance of Carbon-free Hydrazine Fuel Cells Based on Controlled Electrode Structure and Water Management." *Journal of Energy Chemistry* 51: 175 – 181.
[14] Thomassen, M., Borresen, B., Scott, K., Tunold, R. 2006. "A computational Simulation of a Hydrogen/Chlorine Single Fuel Cell." *Journal of Power Sources* 157: 271 – 283.
[15] Babir, F. 2005. "Fuel Cells and Hydrogen Economy." In: *PEM Fuel Cells: Theory and Practice*, edited by Richard C. Dorf 339 - 426, Cambridge: Elsevier Academic Press.
[16] Herrero, E., Feliu, J. M., Aldaz, A. 2003. "*Electrocatalysis, Encyclopedia of Electrchemistry*," edited by A. J. Bard, M. Stratmann, E. J. Calvo 2:443-465 Wiley-VCH Verlag GmbH & Co, Weinheim, Alemania.
[17] Chen, S., Feng X., Weixin, H. 2019. "Surface Chemistry and Catalysis of Oxide Model Catalysts from Single Crystals to Nanocrystals." *Surface Science Reports* 74:100471.
[18] Su, D., Shixue, D., Guoxiu, W. 2014. "Single Crystalline Co_3O_4 Nanocrystals Exposed with Different Crystal Planes for $Li-O_2$ Batteries." *Scientific Reports* 4: 5767.
[19] Cao, S., Feng, F. T., Tang Y., Li, Y., Yu, J. 2016. "Size- and Shape-Dependent Catalytic Performances of Oxidation and Reduction Reactions on Nanocatalysts." *Chemical Society Reviews* 45: 4747-4765.
[20] Feliu, J. M., Herrero, E., Climent V. 2011. "*Electrocatalytic Properties of Stepped Surfaces Catalysis in electrochemistry: from fundamentals to strategies for fuel cell development.*" (E. Santos, W. Schmickler (eds.) 126-164. John Wiley & Sons, Hoboken, EE.UU.

[21] Herrero, E., Feliu J. M. 2017. *"Kinetics at Single Crystal Electrodes Electrochemical Science for a Sustainable Society."* (K. Uosaki (eds.) 113-146, Springer Cham, Suiza.

[22] Solla-Gullón, J., Vidal-Iglesias, F. J., Herrero, E., Feliu, J. M., Aldaz A. 2013. *"Electrocatalysis on Shape-Controlled Pt Nanoparticles Polymer Electrolyte Fuel Cells: Science, Applications, and Challenges"* (A. A. Franco (eds.) 93-152, Pan Stanford Publishing Pte. Ltd, Singapur.

[23] Solla-Gullón, J., Vidal-Iglesias, F. J., Herrero, E., Feliu, J. M., Aldaz A. 2013. *"Electrocatalysis on Shape-Controlled Pt Nanoparticles Polymer Electrolyte Fuel Cells: Science, Applications, and Challenges"* (A. A. Franco (eds.) 93-152. Pan Stanford Publishing Pte. Ltd, Singapur.

[24] Ferreira, P. J., Shao-Horm, Y. 2007. "Formation Mechanism of Pt Single-Crystal Nanoparticles in Proton Exchange Membrane Fuel Cells." *Electrochemical and Solid-State Letters* 10:63-66.

[25] Yu, T., Kim, D. Y., Zhang, H., Xia, Y. 2011. "Platinum Concave Nanocubes with High-Index Facets and Their Enhanced Activity for Oxygen Reduction Reaction." *Angewandte Chemie - International Edition* 50: 2773-2777.

[26] Tiwari, J. N., Tiwari, R. N., Singh, G., Kim, K. 2013. "Recent Progress in the Development of Anode and Cathode Catalysts for Direct Methanol Fuel Cells." *Nano Energy* 2: 553-578.

[27] Sasaki, K., Kuttiyiel, K. A., Adzic, R. R. 2020. "Designing High Performance Pt Monolayer Core-Shell Electrocatalysts for Fuel Cells." *Current Opinion in Electrochemistry* 21: 368-375.

[28] Prokop, M.; Drakselova, M., Bouzek, K. 2020. "Review of the Experimental Study and Prediction of Pt-Based Catalyst Degradation during PEM Fuel Cell Operation." *Current Opinion in Electrochemistry* 20: 20-27.

[29] Zeng, Y., Ji, B., Lv, Z, Zheng, X., Yang, X., Cui, P., Dong, Y., Zhang, X., Jiang, J. 2020. "Rapid Synthesis of Porous Pt-Ni-Cu Coatings with a Wide Composition Range, Tunable Structures and Enhanced

Electrocatalytic Properties." *Journal of Alloys and Compounds* 835:155402.

[30] Viswanathan, B. 2020. "Platinum-Based Anode Catalyst Systems for Direct Methanol Fuel Cells." In *Direct Methanol Fuel Cell Technology,* edited by Kingshuk D, 177-200. Elsevier: Elsevier Inc.

[31] Li, J., Li, L., Wang, M. J., Wang, J., Wei, Z. 2018. "Alloys with Pt-Skin or Pt-Rich Surface for Electrocatalysis." *Current Opinion in Chemical Engineering* 20: 60-67.

[32] Xiao, B. B., Jiang, X. B., Jiang, Q. 2016. "Density Functional Theory Study of Oxygen Reduction Reaction on Pt/Pd$_3$Al(111) Alloy Electrocatalyst." *Physical Chemistry Chemical Physics* 18: 14234-14243.

[33] Shao, M., Chang, Q., Dodelet, J. P., Chenitz, R. 2016. "Recent Advances in Electrocatalysts for Oxygen Reduction Reaction." *Chemical Reviews* 116: 3594-3657.

[34] Li, Y.;Quan, F., Zhu, E., Chen, L., Huang, Y., Chen, C. 2015. "Pt$_x$Cu$_y$ Nanocrystals with Hexa-Pod Morphology and Their Electrocatalytic Performances towards Oxygen Reduction Reaction." *Nano Research* 8: 3342-52.

[35] Zhao, J., Huang, Z., Jian, B., Bai, X., Jian, Q. 2020. "Thermal Performance Enhancement of Air-Cooled Proton Exchange Membrane Fuel Cells by Vapor Chambers." *Energy Conversion and Management* 213:112830.

[36] Açıkkalp, E., Lingen, C., Ahmadi, M. H. 2020. "Comparative Performance Analyses of Molten Carbonate Fuel Cell-Alkali Metal Thermal to Electric Converter and Molten Carbonate Fuel Cell-Thermo-Electric Generator Hybrid Systems." *Energy Reports* 6:10-16.

[37] Jang, J., Kim, D. H., Ahn Mi-K., Min, C-M., Lee, S-B., Byun, J., Pak, C. Lee, J-S. 2020. "Phosphoric Acid Doped Triazole-Containing Cross-Linked Polymer Electrolytes with Enhanced Stability for High-Temperature Proton Exchange Membrane Fuel Cells." *Journal of Membrane Science* 595: 117508.

[38] Hanif, S., Iqbal, N. I., Shi, X., Noor, T., Ali, G., Kannan, A. M. 2020. "NiCo-N-Doped Carbon Nanotubes Based Cathode Catalyst for Alkaline Membrane Fuel Cell." *Renewable Energy* 154:508-16.

[39] Siwal, S. S., Thakur, S., Zhang, Q. B., Thakur, V. K. 2019. "Electrocatalysts for Electrooxidation of Direct Alcohol Fuel Cell: Chemistry and Applications." *Materials Today Chemistry* 14:100182.

[40] Leo, T. J., Raso M. A., Navarro, E., Sánchez-De-La-Blanca, E. 2011. "Comparative Exergy Analysis of Direct Alcohol Fuel Cells Using Fuel Mixtures." *Journal of Power Sources* 196:1178-1183.

[41] Puthiyapura, V. K., Brett, D. J., Russell, A. E., Lin, W. F., Hardacre, C. 2015. "Development of a PtSn bimetallic catalyst for direct fuel cells using bio-butanol fuel." *Chemical Communications*, 51:13412-13415.

[42] Puthiyapura, V. K., Brett, D. J., Russell, A. E., Lin, W. F., Hardacre, C. 2016. "Biobutanol as Fuel for Direct Alcohol Fuel Cells - Investigation of Sn -Modified Pt Catalyst for Butanol Electro-oxidation." *ACS applied materials & interfaces*, 8:12859-12870.

[43] Lamy, C., Belgsir, E. M., Leger, J. M. 2001. "Electrocatalytic oxidation of aliphatic alcohols: application to the direct alcohol fuel cell (DAFC)." *Journal of Applied Electrochemistry*, 31:799-809.

[44] Sandrini, R. M. L. M., Sempionatto, J. R., Tremiliosi-Filho, G., Herrero, E., Feliu, J M., Souza-Garcia, J., Aangelucci, C A. 2019. "Electrocatalytic oxidation of Glycerol on Pt single crystals in alkaline media." *Chem Electro Chem*, 6:4238-4245.

[45] Colmati, F., Tremiliosi-Filho, G., Gonzalez, E. R., Berná, A., Herrero, E., Feliu, J.M. 2009. "The role of the steps in the cleavange of C-C bond during ethanol oxidation on platinum electrodes." *Physical Chemistry Chemical Physics*, 11:9114-9123.

[46] Colmati, F., Antolini, E. 2018. "PtSn electrocatalysts for ethanol oxidation reaction: The effect of metal ratio and thermal treatment" in: *Advanced materials and systems for electrochemical technologies* ed. Mauro Coelho dos Santos, Nova Science Publisher, Inc. New York.

[47] Dürre, P. 2007. "Biobutanol: an attractive biofuel." *Biotechnology Journal: Healthcare Nutrition Technology*, 2:1525-1534.

[48] Takky, D., Beden, B., Leger, J. M., Lamy, C. 1983. "Evidence for the Effect of Molecular Structure on the Electrochemical Reactivity of Alcohols: Part I. Electrooxidation of the Butanol Isomers on Noble Metal Electrodes in Alkaline Medium." *Journal of Electroanalytical Chemistry and Interfacial Electrochemistry* 145:461−466.

[49] Coutinho, J. W. D. Sampaio, A. M. B. S., da Silva, I., Colmati, F., de Lima, R. B. 2018. "Platinum-cadmium electrocatalysts for ethylene glycol electrochemical reaction in perchloric acid electrolyte." *Journal of Solid State Electrochemistry*. 22:3147-3159.

[50] Lamy, C., Lima, A., LeRhun, V., Delime, F., Coutanceau, C., Leger, J. M. 2002. "Recent Advances in the Development of Direct Alcohol Fuel Cells (DAFC)." *Journal of Power Sources*. 105:283−296.

[51] Li, N. H., Sun, S. G., Chen, S. P. 1997. "Studies on the Role of Oxidation States of the Platinum Surface in Electrocatalytic Oxidation of Small Primary Alcohols." *Journal of Electroanalytical Chemistry*. 430:57− 67.

[52] Kim, J. H., Choi, S. M., Nam, S. H., Seo, M. H., Choi, S. H., Kim, W. B. 2008. "Influence of Sn Content on PtSn/C Catalysts for Electrooxidation of C1−C3 Alcohols: Synthesis, Characterization, and Electrocatalytic Activity." *Applied. Catalysis B: Environmental*, 82:89-102.

[53] Takky, D., Beden, B., Leger, J. M., Lamy, C. 1983. "Evidence for the Effect of Molecular Structure on the Electrochemical Reactivity of Alcohols: Part I. Electrooxidation of the Butanol Isomers on Noble Metal Electrodes in Alkaline Medium." *Journal of Electroanalytical Chemistry and Interfacial Electrochemistry* 145:461−466.

[54] Takky, D., Beden, B., Leger, J. M., Lamy, C. 1985. "Evidence for the effect of molecular structure on the electrochemical reactivity of alcohols: Part II. Electrocatalytic oxidation of the butanol isomers on platinum in alkaline medium." *Journal of Electroanalytical Chemistry and Interfacial Electrochemistry*, 193:159-173.

[55] Morin, M. C., Lamy, C., Leger, J. M., Vasquez, J. L., Aldaz, A. 1990. "Structural effects in electrocatalysis: oxidation of ethanol on platinum

single crystal electrodes. Effect of pH." *Journal of Electroanalytical Chemistry and Interfacial Electrochemistry*, 283:287-302.

[56] Takky, D., Beden, B., Leger, J. M., Lamy, C. 1988. "Evidence for the effect of molecular structure on the electrochemical reactivity of alcohols: part III. Electro-oxidation of the butanol isomers on platinum single crystals in an alkaline medium." *Journal of Electroanalytical Chemistry and Interfacial Electrochemistry*, 256:127-136.

[57] Tripković, A. V., Popović, K. D., Lović, J. D. 2001. "The influence of the oxygen-containing species on the electrooxidation of the C1–C4 alcohols at some platinum single crystal surfaces in alkaline solution." *Electrochimica Acta*, 46:3163-3173.

[58] Shao, M. H., Adzic, R. R. 2005. "Electrooxidation of ethanol on a Pt electrode in acid solutions: in situ ATR-SEIRAS study." *Electrochimica Acta*, 50:2415-2422.

[59] dos Reis, R. G., & Colmati, F. 2016. "Electrochemical alcohol oxidation: a comparative study of the behavior of methanol, ethanol, propanol, and butanol on carbon-supported PtSn, PtCu, and Pt nanoparticles." *Journal of Solid State Electrochemistry*, 20:2559-2567.

[60] Busó-Rogero, C., Solla-Gullón, J., Vidal-Iglesias, F. J., Herrero, E., Feliu, J. M. 2016. "Oxidation of ethanol on platinum nanoparticles: suface structure and aggregation effect in alkaline medium." *Journal Solid State Electrochemistry* 20:1095:1106.

[61] Colmati, F., Tremiliosi-Filho, G. Gonzalez, E. R., Berná, A., Herrero, E., Feliu, J. M. 2009. "Surface structure effects on the electrochemical oxidation of ethanol on platinum single crystal electrodes." *Faraday Discussions*. 140:379-397.

[62] Del Colle, V., Souza-Garcia, J., Tremiliosi-Filho, G., Herrero, E.; Feliu, J. M. 2011. "Electrochemical and spectroscopic studies of ethanol oxidation on Pt stepped surfaces modified by tin adatoms." *Physical Chemistry Chemical Physics*, 13:12163-12172.

[63] Colmati, F., Alonso, C. G., Martins, T. D., de Lima, R. B., Ribeiro, A. C. C., Carvalho, de L. L., Sampaio, A. M. B. S., Coutinho J. W. D., de Souza, G. A. Aguiar, L. F., Cordeiro D. S., Costa A. M. F., Ribeiro, T.

S. S., Godoy, P. H. M., de Souza, G. B. M. 2018. "Production of Hydrogen and their Use in Proton Exchange Membrane Fuel Cells" in: *Advances in Hydrogen Generation Technologies*, ed. Murat Eyvaz, Intech Open, London, United Kingdom.

[64] Li, N. Hi; Sun, S. G. 1997. "In situ FTIR spectroscopic studies of the electrooxidation of C4 alcohol on a platinum electrode in acid solutions Part I. Reaction mechanism of 1-butanol oxidation." *Journal of Electroanalytical Chemistry*, 436:65-72.

[65] Puthiyapura, V. K., Brett, D. J., Russel, A. E., Lin, W. F., Hardacre C. 2015. "Preliminari investigation on the electrochemical activity of butanol isomers as potential fuel for direct alcohol fuel cell." *ECS-Transactions* 69:809-816.

BIOGRAPHICAL SKETCHES

Guilhermina Ferreira Teixeira

Affiliation: Federal University of Goiás (UFG)- Institute of Chemistry

Education: PhD, Universidade Estadual Paulista-Instituto de Química-Unesp-IQ, 2015.

Business Address: without number, Esperança Avenue, Samambaia Campus, Zip Code: 74690-900, Goiânia-Goiás- -Brazil

Research and Professional Experience: Area of scientific activities lies in materials chemistry with emphasis in chemical synthesis, characterization and application of multifunctional materials.

Professional Appointments: Post-doctoral Fellow

Publications from the Last 3 Years:

- Teixeira, G. F., Silva Junior, E., Vilela, R., Zaghete, M. A., Colmati, F. 2019. "Perovskite Structure Associated with Precious Metals: Influence on Heterogenous Catalytic Process." *Catalysts* 9: 721.
- Bobic, J. D., Teixeira, G. F., Grigalaitis, R., Gyergyek, S., Petrovi', M. M. Vijatovi', Zaghete, M. A., Stojanovic, B. D. 2019. "PZT-NZF/CF Ferrite Flexible Thick Films: Structural, Dielectric, Ferroelectric, and Magnetic Characterization." *Journal of Advanced Ceramics* 8: 545 – 554.
- Sanches, A. O., Teixeira, G. F., Zaghete, M. A., Malmonge, J. A., Silva, M. J., Sakamoto, W. K. 2019. "Influence of Polymer Insertion on the Dielectric, Piezoelectric and Acoustic Properties of 1-0-3 Polyurethane/Cement-based Piezo Composite." *Materials Research Bulletin* 119: 110541.
- Fernandes, S. L., Gasparotto, G., Teixeira, G. F., Cebim, M. A., Longo, E., Zaghete, M. A. 2018. "Lithium Lanthanum Titanate Perovskite Ionic Conductor: Influence of Europium Doping on Structural and Optical Properties." *Ceramics International* 44: 21578-21584.
- Amoresi, R. A. C., Teodoro, V., Teixeira, G. F., Li, M. S., Simões, A. Z., Perazolli, Longo, E., Zaghete, M. A. 2018. "Electrosteric Colloidal Stabilization for Obtaining $SrTiO_3$/TiO_2 Heterojunction: Microstructural Evolution in the Interface and Photonics Properties." *Journal of the European Ceramic Society* 38: 1621-1631.
- Teixeira, G. F., Silva Junior, E., Simões, A. Z., Longo, E., Zaghete, M. A. 2017. "Unveiling the Correlation between Structural Order-Disorder Character and Photoluminescence Emissions of $NaNbO_3$." *Cryst Eng Comm* 19: 4378.
- Colmati, F., Sgobi, L. F., Teixeira, G. F., Vilela, R. S., Martins, T. D., Figueiredo, G. O. 2019. "Electrochemical Biosensors Containing Pure Enzymes or Crude Extracts as Enzyme Sources for Pesticides and Phenolic Compounds with Pharmacological Property

- Detection and Quantification." In: *Biosensor for Environmental Monotoring*, edited by Toonika Rinken and Kairi Kivirand 367. London, UK: Intech Open Limited.
- Teixeira, G. F., Lustosa, G. M. M. M., Zanetti, S. M., Zaghete, M. A., 2018. "Chemical Synthesis and Epitaxial Growth Methods for Preparation of Ferroelectric Ceramics and Films." In: *Magnetic, Ferroelectric, and Multiferroic Metal Oxides*, edited by Biljana D. Stojanovic 121-137. Amsterdam, Netherlands: Elsevier.
- Lustosa, G. M. M. M., Teixeira, G. F., Bastos, W. B., Zanetti, S. M., Zaghete, M. A. 2018. "Nanosize Ferroelectrics: Preparation, Properties, Application." In: *Magnetic, Ferroelectric, and Multiferroic Metal Oxides*, edited by Biljana D. Stojanovic 139-152. Amsterdam, Netherlands: Elsevier.
- Zaghete, M. A., Perazolli, L. A., Gasparotto, Gisane, Lustosa, G. M. M. M., Biasoto, G., Teixeira, G. F., Jacomaci, N., Amoresi, R. A. C., Fernandes, S, L. 2017. "Multifunctional Complex Oxide Processing." In: *Advances in Complex Functional Materials*, edited by Elson Longo and Felipe de Almeida Laporta 3 - 41. New York,USA: Springer, Cham.
- Lustosa, G. M. M. M., Teixeira, G. F., Costa, J. P. C., Perazolli, L., Zaghete, M. A. 2017. "SnO_2-Thick Films Obtained by Electrophoretic Deposition and Their Technologhical Application." In: *Electrophoretic Deposition (EPD): Advances in Applications and Research*, edited by Nathan Bass 39-66. New York, USA: Nova Science Publisher Inc.
- Teixeira, G. F., Amoresi, R. A. C., Zaghete, M. A., Varela, J. A., Longo, E. 2017. "$NaNbO_3$/PVDF Composite: A Flexible Functional Material." In: *Advances in Solid Oxide Fuel Cells and Electronic Ceramics II: Ceramic Engineering and Science Proceedings, Volume 37, Issue 3, XXXVII*, edited by Mihails Kusnezoff, Narottam P. Bansal, Kiyoshi Shimamura, Manabu Fukushima, and Andrew Gyekenyesi 153-164. New Jersey, USA: John Wiley & Sons, Inc.

Tarso Leandro Bastos

Affiliation: Federal University of Goiás (UFG)- Institute of Chemistry

Education: Bachelor Chemistry, Pontifícia Universidade Católica de Goiás, 2019.

Business Address: without number, Esperança Avenue, Samambaia Campus, Zip Code: 74690-900, Goiânia-Goiás-Brazil

Research and Professional Experience: Area of scientific activities lies in electrochemistry, study of fuel cell and alcohol oxidation by metal alloys electrocatalyst, synthesis, characterization and development of new catalysts and electrodes.

Professional Appointments: Postgraduate Education Fellow

Enrique Herrero

Affiliation: Universidad de Alicante (UA)- Departamento de Química Física Instituto de Electrochimica

Education: PhD. Universidad de Alicante, 1995.

Business Address: Ap 99 E-03080, Alicante, Alicante, España

Research and Professional Experience: His main research field is Surface Electrochemistry and Electrocatalysis. Within this area, he has focused in: Characterization of single crystalline surfaces of noble metals, modified or not by the presence of atomic monolayers; Electrocatalytic activity in model reactions; Oxidation of CO, formic acid, methanol, reduction of O_2, CO_2; Effect of the surface structure of nanoparticles in their electrochemical behavior. He is member of the ISE (International Society of

Electrochemistry). Member of the editorial board of the journals Electrochimica Acta. He has participated in the organizing and scientific committees of congresses organized by the ISE."

Professional Appointments: Professor

Publications from the Last 3 Years:

- Betts, A., Briega-Martos, V., Cuesta, A., and Herrero E. 2020. "Adsorbed Formate is the Last Common Intermediate in the Dual-Path Mechanism of the Electrooxidation of Formic Acid" *ACS Catalysis.* doi.org/10.1021/acscatal.0c00791.
- Briega-Martos, V., Ferre-Vilaplana, A., Herrero, and E., Feliu J. M. 2020. "Why the activity of the hydrogen oxidation reaction on platinum decreases as pH increases" *Electrochimica Acta* doi.org/10.1016/j.electacta.2020.136620.
- Boronat-González, A., Herrero, E., Feliu J. M. 2020. "Determination of the potential of zero charge of Pt/CO electrodes using an impinging jet system." *Journal of Solid State Electrochemistry* doi.org/10.1007/s10008-020-04654-7.
- Gisbert-González, J. M., Cheuquepán, W., Ferre-Vilaplana, A., Herrero, E. et al. 2020. "Citrate adsorption on gold: Understanding the shaping mechanism of nanoparticles." *Journal of Electroanalytical Chemistry*, 114015.
- Briega-Martos, V., Herrero, E., and Feliu J. M. 2020. "Hydrogen peroxide and oxygen reduction studies on Pt stepped surfaces: Surface charge effects and mechanistic consequences." *Electrochimica Acta* 334, 135452.
- Briega-Martos, V., Herrero, E., Feliu, J. M. 2020. "Recent Progress on Oxygen and Hydrogen Peroxide Reduction Reactions on Pt Single Crystal electrodes." *Chinese Journal of Catalysis* 41: 732-738.

- Briega-Martos, V., Herrero, E., and Feliu J.M. 2020. "Recent progress on oxygen and hydrogen peroxide reduction reactions on Pt single crystal electrodes." *Elsevier*.
- Farias, M. J. S., Busó-Rogero, C., Tanaka, A. A., Herrero E., and Feliu, J. M. 2019. "Monitoring of CO Binding Sites on Stepped Pt Single Crystal Electrodes in Alkaline Solutions by in Situ FTIR Spectroscopy." *Langmuir* 36 (3), 704-714.
- Briega-Martos, V., Herrero, E., and Feliu J. M. 2019. "Pt (hkl) surface charge and reactivity." *Current Opinion in Electrochemistry* 17, 97-105.
- Sandrini, R. M. L. M., Sempionatto, J. R., Tremiliosi-Filho, G., Herrero, E., et al. 2019. "Electrocatalytic oxidation of glycerol on platinum single crystals in alkaline media" *Chem Electro Chem* 6 (16), 4238-4245.
- Kamyabi, M. A., Martínez-Hincapié, R., Feliu, J. M., and Herrero E. 2019. "Effects of the Interfacial Structure on the Methanol Oxidation on Platinum Single Crystal Electrodes." *Surfaces* 2 (1), 177-192.
- Briega-Martos, V., Solla-Gullón, J., Koper, M. T. M., Herrero, E., and Feliu, J. M. 2019. "Electrocatalytic enhancement of formic acid oxidation reaction by acetonitrile on well-defined platinum surfaces." *Electrochimica Acta* 295, 835-845.
- Mello, G. A. B., Busó-Rogero, C., Herrero, E., and Feliu, J. M. 2019. "Glycerol electrooxidation on Pd modified Au surfaces in alkaline media: Effect of the deposition method." *The Journal of chemical physics* 150 (4), 041703.
- Ferre-Vilaplana, A., and Herrero, E. 2019. "Why nitrogen favors oxygen reduction on graphitic materials." *Sustainable Energy & Fuels* 3 (9), 2391-2398.
- Briega-Martos, V., Costa-Figueiredo, M., Orts, J. M., Rodes, A., Koper, M. T. M. et al. 2019. "Acetonitrile Adsorption on Pt Single-Crystal Electrodes and Its Effect on Oxygen Reduction Reaction in Acidic and Alkaline Aqueous Solutions." *The Journal of Physical Chemistry C* 123 (4), 2300-2313.

- Busó-Rogero, C., Ferre-Vilaplana, A., Herrero, E., and Feliu J. M. 2019. "The role of formic acid/formate equilibria in the oxidation of formic acid on Pt (111)." *Electrochemistry Communications* 98, 10-14.
- Briega-Martos, V., Mello, G. A. B., Arán-Ais, R. M., Climent, V., Herrero, E., and Feliu, J. M. 2018. "Understandings on the inhibition of oxygen reduction reaction by bromide adsorption on Pt (111) electrodes at different pH values." *Journal of the Electrochemical Society* 165 (15), J3045.
- Gisbert-González, J. M., Feliu, J. M. Ferre-Vilaplana, A., and Herrero, E. 2018. "Why Citrate Shapes Tetrahedral and Octahedral Colloidal Platinum Nanoparticles in Water." *The Journal of Physical Chemistry C* 122 (33), 19004-19014.
- Herrero, E., and Feliu, J. M. 2018. "Understanding formic acid oxidation mechanism on platinum single crystal electrodes." *Current Opinion in Electrochemistry* 9, 145-150.
- Sandrini, R. M. L. M., Sempionatto, J. R., Herrero, E., Feliu, J. M., Souza-Garcia, J. et al. 2018. "Mechanistic aspects of glycerol electrooxidation on Pt (111) electrode in alkaline media." *Electrochemistry Communications* 86, 149-152.
- Arán-Ais, R. M., Solla-Gullón, J., Herrero, E., and Feliu J. M. 2018. "On the quality and stability of preferentially oriented (100) Pt nanoparticles: an electrochemical insight." *Journal of Electroanalytical Chemistry* 808, 433-438.

Juan M. Feliu

Affiliation: Universidad de Alicante (UA)- Departamento de Química Física Instituto de Electrochimica

Education: PhD, Universidad de Barcelona, 1978

Business Address: Ap 99 E-03080, Alicante, Alicante, España

Research and Professional Experience: Research interest deals with the establishment of relationships between surface structure and composition of metallic electrodes and its electrochemical reactivity, within the framework of Surface Electrochemistry and Electrocatalysis. Both aspects are believed to be strongly interconnected, because interfacial properties govern reactivity. To achieve this purpose, single crystal electrodes are prepared and routinely used. The interfacial properties are characterized by using different structure sensitive probes. Determination of interfacial properties and surface stability of well-defined substrates is a key step in this investigation. This methodology has been extended to rationalize the polycrystalline metal/solution interface, including nanoparticles. Surface composition is modified by adsorption of foreign adatoms in a controlled way. The electrocatalytic reactions under scope are those clearly related with the previous, more fundamental approach, and are focused to the surface effects in the kinetics of oxidation/reduction of molecular surface probes, oxidation of potential fuels and small nitrogen-containing molecules, as well as the reduction of oxygen and other green chemistry related species. He has been presidente of the International Society of Electrochemistry (2005-2006).

Professional Appointments: Professor

Publications from the Last 3 Years:

- Briega-Martos, V., Ferre-Vilaplana, A., Herrero, E., and Feliu, J. M. 2020. "Why the Activity of the Hydrogen Oxidation Reaction on Platinum Decreases as pH Increases." *Electrochimica Acta*, 136620.
- Climent, V., and Feliu, J. 2020. "Single Crystal Electrochemistry as an In Situ Analytical Characterization Tool." *Annual Review of Analytical Chemistry* 13: 201-222.
- Boronat-González, A., Herrero, E., and Feliu, J. M. 2020. Determination of the Potential of Zero Charge of Pt/CO Electrodes Using an Impinging Jet System. *Journal of Solid Electrochemistry*. doi.org/10.1007/s10008-020-04654-7.

- Solla-Gullón, J., and Feliu, J. M. 2020. "State-of-the-Art in the Electrochemical Characterization of the Surface Structure of Shape-Controlled Pt, Au and Pd Nanoparticles." *Current Opinion in Electrochemistry* 22: 65-71.
- Fang, Y., Ding, S. Y., Zhang, M., Steinmann, S. N., Hu, R., Mao, B. W, Feliu, J. M., and Tian, Z. K. 2020. "Revisiting the Atomistic Structures at the Interface of Au (111) Electrode–Sulfuric Acid Solution." *Journal of the American Chemical Society* 142: 9439-9446.
- Del Colle, V., Perroni, P. B., Feliu, J. M., Tremiliosi-Filho, G., and Varela, H. 2020. "The Role of Surface Sites on the Oscillatory Oxidation of Methanol on Stepped Pt [n (111)×(110)] Electrodes." *The Journal of Physical Chemistry C* 124: 10993-11004.
- Del Colle, V., Nunes, L. M. S., Angelucci, C. A., Feliu, J. M., and Tremiliosi-Filho, G. 2020. "The Influence of Stepped Pt [n (111)×(110)] Electrodes Towards Glycerol Electrooxidation: Electrochemical and FTIR Studies." *Electrochimica Acta* 346: 136187.
- Gómez-Marín, A. M., Briega-Martos, V., and Feliu, J. M. 2020. "Structure Effects on Electrocatalysts Oxygen Reduction on Te-modified Pt (111) Surfaces: Site-Blocking vs Electronic Effects." *The Journal of Chemical Physics* 152: 134702.
- Sarabia, F. J., Sebastián, P., Climent, V., and Feliu, J, M. 2020. "New Insights into the Pt (hkl)-Alkaline Solution Interphases from the Laser Induced Temperature Jump Method." *Journal of Electroanalytical Chemistry*, 114068.
- Gisbert-González, J. M., Cheuquepán, W., Ferre-Vilaplana, A., Herrero, E., and Feliu, J. M. 2020. "Citrate Adsorption on Gold: Understanding the Shaping Mechanism of Nanoparticles." *Journal of Electroanalytical Chemistry*, 114015.
- Briega-Martos, V., Herrero, E., and Feliu, J. M. 2020. "Hydrogen Peroxide and Oxygen Reduction Studies on Pt Stepped Surfaces: Surface Charge Effects and Mechanistic Consequences." *Electrochimica Acta* 334: 135452.

- Briega-Martos, V., Herrero, E., Feliu, J. M. 2020. "Recent Progress on Oxygen and Hydrogen Peroxide Reduction Reactions on Pt Single Crystal electrodes." *Chinese Journal of Catalysis* 41: 732-738.
- Fang, Y., Dong, J. C., Ding, S. Y., Cheng, J., Feliu, J. M., Li, J. F., and Tian, Z. Q. 2019. "Toward a Quantitative Theoretical Method for Infrared and Raman Spectroscopic Studies on Single-Crystal Electrode/Liquid Interfaces." *Chemical Science* 11: 1425-1430.
- Farias, M. J. S., and Feliu, J. M. 2019. "Determination of Specific Electrocatalytic sites in the oxidation of small molecules on crystalline metal surfaces." *Topics in Current Chemistry* 377: 5.
- Dong, J. C., Su, M., Briega-Martos, V., Li, L., Le, J.-B., Radjenovic, P., Zhou, X. S., Feliu, J. M., Tian, Z. Q., and Li, J. H. F. 2020. "Direct In Situ Raman Spectroscopic Evidence of Oxygen Reduction Reaction Intermediates at High-Index Pt(hkl) Surfaces." *Journal of the American Chemical Society* 142:715-719.
- Farias, M. J. S., Busó-Rogero, C., Tanaka, A. A., Herrero, E., and Feliu, J. M. 2020. "Monitoring of CO Binding Sites on Stepped Pt Single Crystal Electrodes in Alkaline Solutions by in Situ FTIR Spectroscopy." *Langmuir* 36: 704-714.
- Farias, M. J. S., Cheuquepán, W., Tanaka, A. A., and Feliu, J. M. 2020. "Identity of the Most and Least Active Sites for Activation of the Pathways for CO_2 Formation from the Electro-Oxidation of Methanol and Ethanol on Platinum." *ACS Catalysis* 10: 543-555.
- Mantelli, H., Martinez-Hincapie, R., Feliu, J. M., Scherson, D. 2019. "Potential-Induced Acid-Base Chemistry of Adsorbed Species." *Electrochimica Acta* 324: 134793.
- Sarabia, F. J., Climent, V., and Feliu, J. M. 2019. "Interfacial Study of Nickel-Modified Pt (111) Surfaces in Phosphate-Containing Solutions: Effect on the Hydrogen Evolution Reaction." *Chem Phys Chem* 20: 3056-3066.
- Attard, G. A., Souza-Garcia, J., Martínez-Hincapié, R., and Feliu, J. M. 2019. "Nitrate Anion Reduction in Aqueous Perchloric Acid as

- an Electrochemical Probe of Pt {1 1 0}-(1× 1) Terrace Sites." *Journal of Catalysis* 378: 238-247.
- Briega-Martos, V., Herrero, E., and Feliu, J. M. 2019. "Pt (hkl) Surface Charge and Reactivity." *Current Opinion in Electrochemistry* 17: 97-105.
- Santiago, P. V. B., Oliveira, R. A. G., Roquetto, J. M., Akiba, N., Gaubeur, I., Angelucci, C. A., Souza-Garcia, J., and Feliu, J. M. 2019. "Oxide Formation as Probe to Investigate the Competition between Water and Alcohol Molecules for OH Species Adsorbed on Platinum." *Electrochimica Acta* 317: 694-700.
- Björling, A., Carbone, D., Sarabia, F. J., Hammarberg, S., Feliu, J. M., and Solla-Gullón, J., 2019. "Coherent Bragg Imaging of 60 Nm Au Nanoparticles under Electrochemical Control at the Nanomax Beamline." *Journal of synchrotron radiation* 26: 1830-1834.
- Han, Q., Jebaraj, A. J. J., Solla-Gullón, J., Feliu, J., and Scherson, D. 2019. *"Rational Design of Electrocatalytic Interfaces: Cd UPD Mediated Nitrate Reduction on Pd: Au Bimetallic Surfaces."* ECS Meeting Abstract MA2019-02: L02-Electrode Processes 12.
- Sandrini, R. M. L. M., Sempionatto, J. R., Tremiliosi-Filho, G., Herrero, E., Feliu, J. M. Souza-Garcia, J., and Angelucci, C. A. 2019. "Electrocatalytic Oxidation of Glycerol on Platinum Single Crystals in Alkaline Media." *Chem Electro Chem* 6: 4238-4245.
- Martínez-Hincapié, R., Climent, V., and Feliu, J. M. 2019. "Peroxodisulfate Reduction on Platinum Stepped Surfaces Vicinal to the (110) and (100) Poles." *Journal of Electroanalytical Chemistry* 847: 113226.
- Sebastián, P., Gómez, E., Climent, V., and Feliu, J. M. 2019. "Investigating The M (hkl)| Ionic Liquid Interface by Using Laser Induced Temperature Jump Technique." *Electrochimica Acta* 311: 30-40.
- Liu, S., Peng, J., Chen, L., Sebastián, P., Feliu, J. M., Yan, J., and Mao, B. 2019. "In-situ STM and AFM Studies on Electrochemical Interfaces in imidazolium-based ionic liquids." *Electrochimica Acta* 309: 11-17.

- Martínez-Hincapié, R., Climent, V., and Feliu, J. M. 2019. "Investigation of the Interfacial Properties of Platinum Stepped Surfaces Using Peroxodisulfate Reduction as a Local Probe." *Electrochimica Acta* 307: 553-563.
- Flor, A., Feliu, J. M., Tsung, C. K., and Scardi, P. 2019. "Vibrational Properties of Pd Nanocubes." *Nanomaterials (Basel)* 9: 609.
- Martínez-Hincapié, R., Climent, V., and Feliu, J. M. 2019. "New Probes to Surface Free Charge at Electrochemical Interfaces with Platinum Electrodes." *Current Opinion in Electrochemistry* 14: 16-22.
- Kamyabi, M. A., Martínez-Hincapié, R., Feliu, J. M., and Herrero, E. 2019. "Effects of the Interfacial Structure on the Methanol Oxidation on Platinum Single Crystal Electrodes." *Surfaces* 2: 177-192.
- Briega-Martos, V., Solla-Gullón, J., Koper, M. T. M., Herrero, E., and Feliu, J. M. 2019. "Electrocatalytic Enhancement of Formic Acid Oxidation Reaction by Acetonitrile on well-Defined Platinum Surfaces." *Electrochimica Acta* 295: 835-845.
- Pfisterer, J. H. K., Zhumaev, U. E., Cheuquepan, W., Feliu, J. M., and Domke, K. F. 2018. "Stark Effect or Coverage Dependence? Disentangling the EC-SEIRAS Vibrational Shift of Sulfate on Au (111)." *The Journal of chemical physics* 150: 041709.
- Mello, G. A. B., Busó-Rogero, C., Herrero, E., and Feliu, J. M. 2018. "Glycerol Electrooxidation on Pd Modified Au Surfaces in Alkaline Media: Effect of the Deposition Method." *The Journal of chemical physics* 150: 041703.
- Gomez-Marin, A. M., Feliu, J. M., and Ticianelli, E. 2019. "Oxygen Reduction on Platinum Surfaces in Acid Media: Experimental Evidence of a CECE/DISP Initial Reaction Path." *ACS Catalysis* 9: 2238-2251.
- Garnier, E., Vidal-Iglesias, F. J., Feliu, J. M., and Solla-Gullón, J. 2019. "Surface Structure Characterization of Shape and Size Controlled Pd Nanoparticles by Cu UPD: A Quantitative Approach." *Frontiers in chemistry* 7: 527.

- Briega-Martos, V., Costa-Figueiredo, M., Orts, J. M., Rodes, A., Koper, M. T. M. Herrero, E., and Feliu, J. M. 2019. "Acetonitrile Adsorption on Pt Single-Crystal Electrodes and Its Effect on Oxygen Reduction Reaction in Acidic and Alkaline Aqueous Solutions." *The Journal of Physical Chemistry C* 123: 2300-2313.
- Dong, J. C., Zhang, X. G., Briega-Martos, V., Jin, X., Yang, J., Chen, S., Yang, Z. L., Wu, D. Y., Feliu, J. M., and Williams, C. T., Tian, Z. Q., Li, J. F. 2019. "In Situ Raman Spectroscopic Evidence For Oxygen Reduction Reaction Intermediates At Platinum Single-Crystal Surfaces." *Nature Energy* 4: 60-67.
- Busó-Rogero, C., Ferre-Vilaplana, A., Herrero, E., and Feliu, J. M. 2019. "The Role of Formic Acid/Formate Equilibria in the Oxidation of Formic Acid on Pt (111)." *Electrochemistry Communications* 98: 10-14.
- Sarabia, F. J., Sebastián-Pascual, P., Koper, M. T. M., Climent, V., and Feliu, J. M. 2019. "Effect of the Interfacial Water Structure on the Hydrogen Evolution Reaction on Pt (111) Modified with Different Nickel Hydroxide Coverages in Alkaline Media." *ACS applied materials & interfaces* 11: 613-623.
- Farias, M. J. S., Cheuquepán, W., Tanaka, A. A., and Feliu, J. M. 2018. "Requirement of Initial Long-Range Substrate Structure in Unusual CO Pre-Oxidation on Pt (111) Electrodes." *Electrochemistry Communications* 97: 60-63.
- Chumillas, S., Palomäki, T., Zhang, M., Laurila, T., Climent, V., and Feliu, J. M. 2018. "Analysis of Catechol, 4-Methylcatechol and Dopamine Electrochemical Reactions on Different Substrate Materials and pH Conditions." *Electrochimica Acta* 292: 309-321.
- Moglianetti, M., Solla-Gullón, J., Donati, P., Pedone, D., Debellis, D., Sibillano, T., Brescia, R., Giannini, C., Montiel, V., Feliu, J. M., and Pompa, P. P. 2018. "Citrate-Coated, Size-Tunable Octahedral Platinum Nanocrystals: A Novel Route for Advanced Electrocatalysts." *ACS applied materials & interfaces* 10: 41608-41617.

- Briega-Martos, V., Mello, G. A. B., Arán-Ais, R. M., Climent, V., Herrero, E., and Feliu, J. M. 2018. "Understandings on the Inhibition of Oxygen Reduction Reaction by Bromide Adsorption on Pt (111) Electrodes at Different pH Values." *Journal of the Electrochemical Society* 165: J3045.
- Chumillas, S., Maestro, B., Feliu, J. M., and Climent, V. 2018. "Comprehensive Study of the Enzymatic Catalysis of the Electrochemical Oxygen Reduction Reaction (ORR) by Immobilized Copper Efflux Oxidase (CueO) From Escherichia coli." *Frontiers in Chemistry* 6: 358.
- Sebastián, P., Tułodziecki, M., Bernicola, M. P., Climent, V., Gómez, E., Shao-Horn, Y., and Feliu, J. M. 2018. "Use of CO as a Cleaning Tool of Highly Active Surfaces in Contact with Ionic Liquids: Ni Deposition on Pt (111) Surfaces in IL." *ACS Applied Energy Materials* 1: 4617-4625.
- Cheuquepán, W., Rodes, A., Orts, J. M., and Feliu, J. M. 2018. "Spectroelectrochemical and Density Functional Theory Study of Squaric Acid Adsorption and Oxidation at Gold Thin Film and Single Crystal Electrodes." *The Journal of Physical Chemistry C* 122: 22352-22365.
- Chumillas, S., Maestro, B., Feliu, J. M., and Climent, V. 2018. "Optimal conditions for the Electrochemical Oxygen Reduction Reaction (ORR) promoted by immobilized CueO from E. Coli." *Frontiers in Chemistry* 6: id.358.
- Gisbert-González, J. M., Feliu, J. M., Ferre-Vilaplana, A., and Herrero, E. 2018. "Why Citrate Shapes Tetrahedral and Octahedral Colloidal Platinum Nanoparticles in Water." *The Journal of Physical Chemistry C* 122: 19004-19014.
- Mello, G. A. B., Briega-Martos, V., Climent, V., and Feliu, J. M. 2018. "Bromide Adsorption on Pt (111) over a Wide Range of pH: Cyclic Voltammetry and CO Displacement Experiments." *The Journal of Physical Chemistry C* 122:18562-18569.
- Gómez-Marín, A., Feliu, J. M., and Ticianelli, E. 2018. "Reaction Mechanism for Oxygen Reduction on Platinum: Existence of a Fast

Initial Chemical Step and a Soluble Species Different from H_2O_2." *ACS Catalysis* 8: 7931-7943.
- Molodkina, E. B., Danilov, A. I., Ehrenburg, M. R., and Feliu, J. M. 2018. "Regularities of Nitrate Electroreduction on Pt (S)[N (100) X (110)] Stepped Platinum Single Crystals Modified by Copper Adatoms." *Electrochimica Acta* 278: 165-175.
- Sarabia, F. J., Climent, V., and Feliu, J. M. 2018. "Underpotential Deposition of Nickel on Platinum Single Crystal Electrodes." *Journal of Electroanalytical Chemistry* 819: 391-400.
- Gomez-Marin, A. M., and Feliu, J. M. 2018. "Oxygen Reduction at platinum Electrodes: The Interplay between Surface and Surroundings Properties." *Current Opinion in Electrochemistry* 9: 166-172.
- Herrero, E., and Feliu, J. M. 2018. "Understanding Formic Acid Oxidation Mechanism on Platinum Single Crystal Electrodes." *Current Opinion in Electrochemistry* 9: 145-150.
- González-Arribas, E., Falk, M., Aleksejeva, O., Bushnev, S., Sebastián, P. Feliu, J. M., and Shleev, S. 2018. "A Conventional Symmetric Biosupercapacitor Based n on Modified Gold Electrodes." *Journal of Electroanalytical Chemistry* 816: 253-258.
- Feliu, J. M. 2018. Local Probes at the Pt (hkl)/Electrolyte Interface. *Abstracts of Papers of the American Chemical Society 255*.
- Martínez-Hincapié, R., Climent, V., and Feliu, J. M. 2018. "Peroxodisulfate Reduction as a Probe to Interfacial Charge." *Electrochemistry Communications* 88: 43-46.
- Rizo, R., Arán-Ais, R. M., Padgett, E., Muller, D. A., Lázaro, M. J., Solla-Gullón, J., Feliu, J. M., Pastor, E., and Abruña, H. D. 2018. "Pt-Rich$_{core}$/Sn-Rich$_{subsurface}$/Pt$_{skin}$ Nanocubes as Highly Active and Stable Electrocatalysts for the Ethanol Oxidation Reaction." *Journal of the American Chemical Society* 140: 3791-3797.
- Sebastián, P., Giannotti, M. I., Gómez, E., and Feliu, J. M., 2018. "Surface Sensitive Nickel Electrodeposition in Deep Eutectic Solvent." *ACS Applied Energy Materials* 1: 1016-1028.

- Farias, M. J. S., Cheuquepán, W., Tanaka, A.A., and Feliu, J. M. 2018. "Unraveling the Nature of Active Sites in Ethanol Electrooxidation by Site-Specific Marking of a Pt Catalyst with Isotope-Labeled ^{13}CO." *The Journal of Physical Chemistry Letters* 9: 1206-1210.
- Isoaho, N., Sainio, S., Wester, N., Botello, L., Johansson, L. S., Peltola, E. Climent, V., Feliu, J. M., Koskinen, J., and Laurila, T. 2018. "Pt-grown Carbon Nanofibers for Detection of Hydrogen Peroxide." *RSC advances* 8: 12742-12751.
- Sebastián, P., Climent, V., Feliu, J. M., and Gómez, E. 2018. "Ionic Liquids in the Field of Metal Electrodeposition." In: *Encyclopedia of Interfacial Chemistry: Surface Science and Electrochemistry* 2018, 690-700.
- Sandrini, R. M. L. M., Sempionatto, J. M., Herrero, E., Feliu, J. M., Souza-Garcia, J., and Angelucci, C. A. 2018. "Mechanistic Aspects of Glycerol Electrooxidation on Pt (111) Electrode in Alkaline Media." *Electrochemistry Communications* 86: 149-152.
- Sitta, E., Silva, K. N., and Feliu, J. M. 2017. "Hydrogen Peroxide Oxidation/Reduction Reaction on Platinum Surfaces." In: *Encyclopedia of Interfacial Chemistry: Surface Science and Electrochemistry*, edited by Klaus Wandelt 682-689. Amsterdam, Netherlands: Elsevier.
- Gómez-Marín, A. M., and Feliu, J. M. 2017. "Oxygen Reduction on Platinum Single Crystal Electrodes." *Encyclopedia of Interfacial Chemistry: Surface Science and Electrochemistry 2018*, 820-830.
- Arán-Ais, R. M., Solla-Gullón, J., Herrero, E., and Feliu, J. M. 2018. "On The Quality and Stability of Preferentially Oriented (100) Pt Nanoparticles: an electrochemical insight." *Journal of Electroanalytical Chemistry* 808: 433-438.
- Briega-Martos, V., Herrero, E., and Feliu, J. M. 2017. "The Inhibition of Hydrogen Peroxide Reduction at Low Potentials on Pt (111): Hydrogen Adsorption or Interfacial Charge?" *Electrochemistry Communications* 85: 32-35.

- Isoaho, N., Wester, N., Peltola, E., Johansson, L. S., Boronat, A., Koskinen, J., Feliu, J. M., Climent, V., and Laurila, T. 2017. "Amorphous Carbon Thin Film Electrodes with Intrinsic Pt-Gradient for Hydrogen Peroxide Detection." *Electrochimica Acta* 251: 60-70.
- Jukk, K., Kongi, N., Tammeveski, K., Arán-Ais, R. M., Solla-Gullón, J., and Feliu, J. M. 2017. "Loading Effect of Carbon-Supported Platinum Nanocubes on Oxygen Electroreduction." *Electrochimica Acta* 251: 155-166.
- Gómez-Marín, A. M., Boronat, A., and Feliu, J. M. 2017. "Electrocatalytic Oxidation and Reduction of H_2O_2 on Au Single Crystals." *Russian Journal of Electrochemistry* 53: 1029-1041.
- Cheuquepán, W., Rodes, A., Orts, J. M., and Feliu, J. M. 2017. "Spectroelectrochemical Detection of Specifically Adsorbed Cyanurate Anions at Gold Electrodes with (111) Orientation in Contact with Cyanate and Cyanuric Acid Neutral Solutions." *Journal of Electroanalytical Chemistry* 800: 167-175.
- Boronat-González, A., Herrero, E., and Feliu, J. M. 2017. "Heterogeneous Electrocatalysis of Formic Acid Oxidation on Platinum Single Crystal Electrodes." *Current Opinion in Electrochemistry* 4: 26-31.
- Acevedo, R., Poventud-Estrada, C. M., Morales-Navas, C., Martínez-Rodríguez, R. A., Ortiz-Quiles, E., Vidal-Iglesias, F. J., Sollá-Gullón, J., Nicolau, E., Feliu, J. M., Echegoyen, L., and Cabrera, C. R. 2017. "Chronoamperometric Study of Ammonia Oxidation in a Direct Ammonia Alkaline Fuel Cell under the Influence of Microgravity." *Microgravity Science and Technology* 29: 253-261.
- Briega-Martos, V., Herrero, E., and Feliu, J. M. 2017. "Effect of pH and Water Structure on the oxygen Reduction Reaction on Platinum Electrodes." *Electrochimica Acta* 241: 497-509.
- Schäfer, P., Lalitha, A., Sebastian, P., Meena, S. K, Feliu, J. M., Sulpizi, M., van der Veen, M. A., and Domke, K. F. 2017. "Trimesic acid on Cu in ethanol: Potential-dependent Transition from 2-D

- Adsorbate to 3-D Metal-Organic Framework." *Journal of Electroanalytical Chemistry* 793: 226-234.
- Arán-Ais, R. M., Vidal-Iglesias, F. J., Farias, M. J. S., Solla-Gullón, J., Montiel, V., Herrero, E., and Feliu, J, M. 2017. "Understanding CO Oxidation Reaction on Platinum Nanoparticles." *Journal of Electroanalytical Chemistry* 793: 126-136.
- Attard, G. A., Hunter, K., Wright, E., Sharman, J., Martínez-Hincapié, R., and Feliu, J. M. 2017. "The Voltammetry of Surfaces Vicinal to Pt {110}: Structural Complexity Simplified by CO Cooling." *Journal of Electroanalytical Chemistry* 793: 137-146.
- Farias, M. J. S., Cheuquepán, W., Tanaka, A. A., and Feliu, J. M. 2017. "Nonuniform Synergistic Effect of Sn and Ru in Site-Specific Catalytic Activity of Pt at Bimetallic Surfaces toward CO Electro-Oxidation." *ACS Catalysis* 7: 3434-3445.
- Sebastián, P., Gómez, E., Climent, V., and Feliu, J. M. 2017. "Copper Underpotential Deposition at Gold Surfaces in Contact with a Deep Eutectic Solvent: New insights." *Electrochemistry Communications* 78: 51-55.
- Gisbert, R., Boronat-González, A., Feliu, J. M., and Herrero, E. 2017. "The Role of Adsorption in the Electrocatalysis of Hydrazine on Platinum Electrodes." *Chem Electro Chem* 4: 1130-1134.
- Ganassin, A., Sebastián, P., Climent, V., Schuhmann, W., Bandarenka, A. S., and Feliu, J. 2017. "On the pH Dependence of the Potential of Maximum Entropy of Ir (111) Electrodes." *Scientific Reports* 7: 1-14.
- Martinez-Rodriguez, R., Vidal-Iglesias, F., Solla-Gullon, J., Cabrera, C., and Feliu, J. 2017. "Electrochemical Behavior of Shape Controlled Pt-Rh Nanoparticles for Ammonia Oxidation in Alkaline Medium for Direct Alkaline FuelCell Application." *Abstracts of Papers of the American Chemical Society* 253.
- Ledezma-Yanez, I., Wallace, W. D. Z., Sebastián-Pascual, P., Climent, V., Feliu, J. M., and Koper, M. T. M. 2017. "Interfacial Water Reorganization as a pH-dependent Descriptor of the

Hydrogen Evolution Rate on Platinum Electrodes." *Nature Energy* 2: 1-7.

- Climent, V., and Feliu, J. M. 2017. "Surface electrochemistry with Pt single-crystal electrodes." In: *Advances in Electrochemical Science and Engineering*: *Nanopatterned and Nanoparticle-Modified Electrodes*, edited by Richard C. Alkire, Philip N. Bartlett, Jacek Lipkowski 17: 1-57. Weinheim, Germany: Wiley-VCH Verlag GmbH & Co.
- Martínez-Hincapié, R., Sebastián-Pascual, P., Climent, V., and Feliu, J. M. 2017. "Investigating Interfacial Parameters with Platinum Single Crystal Electrodes." *Russian Journal of Electrochemistry* 53: 227-236.
- Cheuquepán, W., Orts, J. M., Rodes, A., and Feliu, J. M. 2017. "Voltammetric and in Situ Infrared Spectroscopy Studies of Hydroxyurea Electrooxidation at Au (111) Electrodes in $HClO_4$ Solutions." *Electrochemistry Communications* 76: 34-37.
- Sebastián, P., Martínez-Hincapié, R., Climent, V., and Feliu, J. M. 2017. "Study of the Pt (111)| Electrolyte Interface in the Region Close to Neutral pH Solutions by the Laser Induced Temperature Jump Technique." *Electrochimica Acta* 228: 667-676.
- Briega-Martos, V., Ferre-Vilaplana, A., de la Peña, A., Segura, J. L., Zamora, L. Feliu, J. M., and Herrero, E. 2017. "An Aza-fused π-Conjugated Microporous Framework Catalyzes the Production of Hydrogen Peroxide." *ACS Catalysis* 7: 1015-1024.
- Farias, M. J. S., Busó-Rogero, C.,Vidal-Iglesias, F. J., Solla-Gullón, J., Solla-Gullón, J., Camara, G. A., and Feliu, J. M. 2017. "Mobility and Oxidation of Adsorbed CO on Shape-Controlled Pt Nanoparticles in Acidic Medium." *Langmuir* 33: 865-871.
- Ferre-Vilaplana, A., Perales-Rondón, J. V., Busó-Rogero, C., Feliu, J. M., and Herrero, E. 2017. "Formic Acid Oxidation on Platinum Electrodes: A Detailed Mechanism Supported by Experiments and Calculations on Well-Defined Surfaces." *Journal of Materials Chemistry A* 5: 21773-21784.

- Herrero, E., Climent, V., and Feliu, J. M. 2017. "Antonio Aldaz Riera (1943-2015)." *Journal of Electroanalytical Chemistry* 793: 1-7.
- Herrero, E., and Feliu, J. M. 2017. "Kinetics at Single Crystal Electrodes." In: *Electrochemical Science for a Sustainable Society. A Tribute to John O'M Bockris*, edited by Kohei Uosaki 113-146. New York, USA: Springer Nature.
- Farias, M. J. S., Mello, G. A. B.,Tanaka, A. A., and Feliu, J. M. 2017. "Site-specific Catalytic Activity of Model Platinum Surfaces in Different Electrolytic Environments as Monitored by the CO Oxidation Reaction." *Journal of Catalysis* 345: 216-227.

Flavio Colmati

Affiliation: Federal University of Goiás (UFG)- Institute of Chemistry

Education: PhD, Sao Carlos Institute of Chemistry, University of Sao Paulo, 2007.

Business Address: without number, Esperança Avenue, Samambaia Campus, Goiânia-Goiás- Brazil

Research and Professional Experience: Electrocatalysts, Proton Exchange Membrane Fuel cell, Direct Ethanol Fuel Cell, Ethanol Oxidation Reaction, Oxygen Reduction Reaction and Pt Single Crystals.

Professional Appointments: Professor

Publications from the Last 3 Years:

- Teixeira, G. F., Silva Junior, E., Vilela, R., Zaghete, M. A., Colmati, F. 2019. "Perovskite Structure Associated with Precious Metals: Influence on Heterogenous Catalytic Process." *Catalysts* 9: 721.

- Colmati, F., Magalhães, M. M., Souza, R., Ciaprina, E. G., Gonzalez, E. R. 2019. "Direct Ethanol Fuel Cells: The Influence of Structural and Electronic Effects on Pt-Sn/C Electrocatalysts." *International Journal of Hydrogen Energy* 44: 28812-28820.
- Oliveira, V. X. G., Dias, A. A., Carvalho, L. L., Cardoso, T. M. G., Colmati, F., Coltro, W. K. T. 2018. "Determination of Ascorbic Acid in Commercial Tablets Using Pencil Drawn Electrochemical Paper-based Analytical Devices." *Analytical Sciences* 34: 91-95.
- Benjamin, S. R., Vilela, R. S., Camargo, H. S., Guedes, M. I. F., Fernandes, K. F., Colmati, F. 2018. "Enzymatic Electrochemical Biosensor Based on Multiwall Carbon Nanotubes and Cerium Dioxide Nanoparticles for Rutin Detection." *International Journal of Electrochemical Science* 13: 563-586.
- Siqueira Leite, K. C., Garcia, L. F., Sanz, G., Colmati, F., Souza, A. R., Costa Batista, D., Menegatti, R., Souza Gil, E., Luque, R. 2018. "Electrochemical Characterization of a Novel Nimesulide Anti-Inflammatory Drug Analog: LQFM-091." *Journal of Electroanalytical Chemistry* 818: 92-96.
- Carvalho, L. L., Tanaka, A. A., Colmati, F. 2018. "Palladium-Platinum Electrocatalysts for the Ethanol Oxidation Reaction: Comparison of Electrochemical Activities in Acid and Alkaline Media." *Journal of Solid State Electrochemistry* 22: 1471-1481.
- Coutinho, J. W. D., Sampaio, A. M. B. S., Silva, I. S., Colmati, F., Lima, R. B. 2018. "Platinum-Cadmium Electrocatalyst for Ethylene Glycol Electrochemical Reaction in Perchloric Acid Electrolyte." *Journal of Solid State Electrochemistry* 22: 3147-3159.
- Carvalho, L. L., Colmati, F., Tanaka, A. A., 2017. "Nickel-Palladium Electrocatalysts for Methanol, Ethanol, and Glycerol Oxidation Reactions." *International Journal of Hydrogen Energy* 42: 16118-16126.
- Santos, G. P. N., Vilela, R. S., Benjamin, S. R., Fernandes, K. F., Colmati, F. 2020. "Enzymatic Electrochemical Biosensors Based on Carbon Paste Electrodes Modified with Polyphenol Oxidase Enzymes for Rutin Detection." In: *An Essential Guide to Rutin,*

- edited by Joachim E. Bach 61-80. New York, USA: Nova Science Publisher Inc.
- Colmati, F., Sgobi, L. F., Teixeira, G. F., Vilela, R. S., Martins, T. D., Figueiredo, G. O. 2019. "Electrochemical Biosensors Containing Pure Enzymes or Crude Extracts as Enzyme Sources for Pesticides and Phenolic Compounds with Pharmacological Property Detection and Quantification." In: *Biosensor for Environmental Monitoring*, edited by Toonika Rinken and Kairi Kivirand 367. London, UK: IntechOpen Limited.
- Linares, J. J., Vieira, C. C., Costa Santos, J. B., Magalhães, M. M., Santos, J. R. N., Carvalho, L. L. Reis, R. G. C. S., Colmati, F. 2019. "Electrochemical Reforming of Alcohols." In: *Energy and Environment Series*, edited by Keith Scott 94-135. Cambridge, UK: Royal Society of Chemistry.
- Souza, G. A., Martins, T. D., Colmati, F. 2019. "Dynamic Luminescent Biosensors Based on Peptides for Oxygen Determination. In: *Biosensors for Environmental Monitoring*, edited by Toonika Rinken and Kairi Kivirand 10. London, UK: InTechOpen Limited.
- Colmati, F., Antolini, E. 2018. "PtSn Electrocatalysts for Ethanol Oxidation Reaction: The Effect of Metal Ratio and Thermal Treatment." In: *Advanced Materials and Systems for Electrochemical Technologies*, edited by Mauro Coelho dos Santos 3-26. New York, USA: Nova Science Publishers.
- Martins, T. D., Ribeiro, A. C. C., Souza, G. A., Cordeiro, D. S., Silva, R. M., Colmati, F., Lima, R. B., Aguiar, L. F., Carvalho, L. L. Reis, R. G. S., Santos, W. D., 2018. "New Materials to Solve Energy Issues through Photochemical and Photophysical Processes: The Kinetics Involved." In: *Advanced Chemical Kinetics*, edited by Muhammad A. Farrukh 1. London, UK: InTechOpen Limited.
- Colmati, F., Alonso, C. G., Martins, T. D., Lima, R. B., Ribeiro, A. C. C., Carvalho, L. L., Souza, G. A., Aguiar, L. F., Cordeiro, D. S., Costa, A. M. F., Ribeiro, T. S. S., Godoy, P. H. M., Souza, G. B. M., .2018. "Production of Hydrogen and their Use in Proton Exchange

Membrane Fuel Cells." In: *Advances in Hydrogen Generation Technologies*, edited by Murat Eyvaz 63-78. London, UK: IntechOpen Limited.

INDEX

A

ABE, viii, 2, 11, 12, 14, 23, 28, 31, 35, 42, 43, 44, 45, 46, 47, 49, 50, 51, 52, 53, 54, 55, 56, 57, 58, 59, 60, 61, 62, 63, 64, 65, 66, 68, 71, 72, 73, 74, 75, 78, 79, 87, 94
ABE fermentation, viii, 2, 14, 28, 31, 45, 49, 50, 54, 57, 59, 60, 61, 63, 65, 66, 73, 74, 75, 87
access, 30, 31, 32, 33
accessibility, 36, 37, 38, 41
acetaldehyde, 3, 9, 16, 110, 116, 117
acetic acid, 37, 38, 43, 44, 110, 116, 117
acetone, vii, viii, 2, 10, 12, 21, 23, 28, 38, 45, 47, 48, 52, 53, 55, 56, 58, 60, 66, 68, 69, 70, 71, 72, 74, 75, 76, 77, 79, 83, 84, 85, 86, 94
acid, 8, 17, 20, 37, 38, 40, 42, 43, 44, 45, 51, 56, 57, 67, 72, 79, 84, 85, 91, 92, 96, 109, 110, 112, 114, 116, 117, 118, 119, 126, 127, 128, 131, 133, 134, 144
active site, ix, 102, 106, 116
adsorption, 17, 53, 56, 58, 59, 68, 70, 75, 78, 79, 80, 81, 85, 105, 106, 107, 108, 114, 118, 120, 132, 133, 134, 135, 136, 140, 141, 143, 145

agro-industrial residues, 33
alcohols, 3, 9, 18, 22, 31, 32, 33, 39, 63, 66, 70, 103, 104, 109, 110, 115, 116, 125, 126, 127
alkaline media, 125, 133, 134
alternative energy, vii, viii, 2
ammonia, 38, 40, 41
atoms, 7, 105, 107, 108
ATP, 2, 12, 47

B

bacteria, 11, 16, 24, 41, 42, 45, 59, 60, 77, 79
base, ix, 4, 19, 43, 89, 102
biobutanol/bio-butanol, vii, viii, 4, 6, 18, 19, 23, 24, 28, 30, 31, 33, 34, 35, 52, 54, 55, 56, 57, 59, 60, 61, 62, 64, 65, 68, 70, 71, 73, 75, 76, 80, 81, 82, 84, 89, 90, 92, 96, 97, 125
biodiesel, 18, 21, 29, 87
biofuel(s), 2, 3, 4, 18, 20, 23, 24, 28, 29, 36, 61, 65, 68, 69, 70, 72, 73, 74, 79, 82, 84, 85, 87, 125
biomass, v, vii, viii, 1, 4, 11, 16, 17, 23, 27, 28, 29, 34, 35, 36, 37, 39, 40, 41, 42, 45,

50, 51, 59, 61, 62, 67, 68, 69, 70, 73, 74, 77, 78, 79, 82, 83, 84, 85, 94, 95, 110
biorefinery, 22, 30, 76, 86, 94, 95
biotechnology, 21, 23, 24
Brazil, 60, 101, 128, 131, 147
brevis, 47, 68, 85
butanol, v, vii, viii, ix, 1, 2, 3, 4, 5, 6, 7, 8, 9, 10, 11, 12, 13, 14, 15, 16, 17, 18, 19, 20, 21, 22, 23, 24, 27, 30, 31, 33, 45, 47, 48, 49, 52, 53, 54, 59, 60, 61, 62, 63, 65, 66, 67, 68, 69, 70, 71, 72, 73, 74, 75, 76, 77, 78, 79, 80, 81, 82, 83, 84, 85, 86, 87, 88, 90, 91, 93, 94, 96, 101, 102, 109, 110, 111, 112, 113, 114, 115, 116, 118, 119, 120, 125, 126, 127, 128
butanol electrooxidation, 102
butanol oxidation reaction, v, vii, ix, 101, 102, 112, 118, 120

C

candidates, 31, 34, 104
carbohydrates, 7, 11, 30, 38, 47, 49, 61
carbon, 3, 4, 5, 6, 7, 16, 18, 20, 29, 70, 112, 115, 116, 117, 119, 122, 127
carbon dioxide, 29, 117, 119
carbon monoxide, 4, 116, 117
catalyst, ix, 3, 16, 38, 40, 102, 105, 106, 107, 108, 109, 114, 115, 116, 125
catalytic activity, 105, 108, 120
C-C, 110, 117, 118, 125
cellulose, 11, 19, 33, 34, 35, 37, 38, 39, 40, 41, 42, 72, 78
ceramic(s), 75, 129, 130
challenges, viii, 23, 28, 30, 61, 70, 82
chemical(s), vii, ix, 1, 3, 6, 7, 11, 18, 19, 22, 24, 29, 35, 37, 38, 39, 40, 68, 69, 70, 82, 89, 92, 97, 102, 103, 110, 116, 119, 120, 128, 133, 139
China, 30, 60, 77
chromatography, 15, 25, 117

classification, 35, 39, 42
clean energy, 103, 109, 120
cleavage, 110, 117, 118
Clostridium, 2, 11, 14, 15, 16, 21, 23, 24, 42, 43, 44, 45, 46, 47, 48, 62, 66, 67, 68, 69, 71, 72, 73, 74, 75, 76, 77, 78, 79, 83, 85
CO_2, 39, 40, 51, 110, 115, 116, 118, 119, 131, 137
combustion, viii, ix, 3, 18, 21, 22, 27, 63, 101, 103
commercial, 10, 20, 42, 78
composition, 30, 33, 34, 105, 108, 120, 135
compounds, ix, 4, 19, 38, 40, 41, 42, 43, 44, 45, 54, 101, 104
consumption, 3, 20, 47, 49, 119
corn steep liquor (CSL), viii, 2
cosmetic(s), viii, 2, 4, 8, 19, 20
cost, 16, 20, 35, 37, 38, 39, 40, 41, 49, 58, 59, 61, 64, 66
crop(s), 29, 31, 33, 76, 77
crystalline, 37, 103, 106, 108, 112, 131, 137
crystallinity, 35, 36, 37, 38, 39
culture, 13, 15, 21, 23, 51, 66, 67, 84

D

DAFC, 102, 125, 126
degradation, 38, 40, 41, 42, 71, 79
dehydration, 72, 90, 118
desorption, 56, 59, 80, 106
destruction, 17, 29, 59
direct alcohol fuel cell, 102, 109, 110, 119, 125, 126, 128
direct piezoelectric effect, 103
distillation, 10, 17, 48, 53, 54, 55, 58, 59, 63, 64, 68, 70, 72, 76, 80, 83, 87, 88, 89, 90, 91, 98, 99
DSM, 43, 44, 52, 77

E

economics, 20, 66, 76, 79, 84
electricity, ix, 101, 103, 109, 110, 115, 120
electrochemical oxidation of butanol, 111
electrochemistry, 110, 122, 131, 146
electrodes, 103, 104, 106, 110, 111, 112, 113, 114, 116, 118, 120, 125, 127, 131, 132, 133, 134, 135, 137, 146
electrolyte, 104, 120, 121, 126
electron(s), 37, 104, 110, 116, 120
energy, vii, ix, 1, 5, 28, 35, 36, 37, 38, 48, 52, 54, 56, 57, 59, 60, 62, 64, 66, 83, 84, 87, 89, 93, 97, 101, 102, 103, 104, 105, 107, 108, 109, 110, 112, 115, 119, 120
energy consumption, 28, 35, 37, 52, 57, 64, 66
energy density, viii, 2, 110
energy efficiency, 64, 84, 115
engineering, 17, 18, 22, 23, 24, 67, 73, 77, 83, 89, 97
environment(s), ix, 28, 38, 40, 41, 59, 65, 101
environmental impact, viii, 28, 64, 66, 102
enzyme(s), viii, 2, 12, 15, 23, 39, 41, 42, 62, 67, 79, 80, 85
equipment, 37, 40, 62, 64
ethanol, vii, viii, 1, 2, 5, 12, 15, 16, 18, 21, 22, 23, 27, 29, 30, 31, 33, 34, 38, 44, 45, 47, 48, 52, 53, 55, 56, 58, 60, 63, 65, 66, 67, 68, 69, 70, 71, 72, 74, 75, 77, 78, 79, 80, 82, 83, 84, 85, 87, 88, 91, 94, 109, 110, 112, 114, 115, 116, 125, 126, 127, 144
evolution, vii, 1, 119
extraction, 14, 17, 50, 53, 54, 55, 56, 57, 58, 63, 64, 70, 74

F

feedstock(s), 29, 30, 69, 72, 77, 81, 84, 88, 96
fermentation, vii, viii, 2, 4, 7, 11, 12, 13, 14, 15, 16, 17, 19, 23, 24, 28, 30, 31, 38, 40, 41, 43, 45, 46, 48, 49, 50, 51, 52, 54, 57, 59, 60, 61, 62, 63, 64, 66, 68, 69, 71, 72, 73, 74, 75, 76, 77, 78, 79, 81, 82, 83, 84, 85, 86, 87, 91, 93, 94
fiber, 37, 40, 43, 49, 74, 79
fluid, 8, 18, 19, 40, 55, 56, 67
food, vii, 1, 6, 7, 11, 14, 20, 25, 29, 32, 76, 77, 102, 110
formation, 8, 10, 11, 12, 19, 20, 23, 24, 40, 54, 58, 68, 71, 72, 104, 110, 116, 117, 118, 119
formula, 6, 7, 8, 10
fruits, 34, 61, 84
FTIR, 118, 128, 133, 136, 137
fuel cell, vii, ix, 101, 102, 103, 104, 106, 107, 108, 109, 110, 111, 112, 113, 114, 115, 116, 117, 119, 120, 121, 122, 123, 124, 125, 128, 130, 131, 148, 150
fuel cells, vii, ix, 101, 103, 107, 108, 109, 110, 111, 112, 115, 120, 121, 122, 123, 124, 128, 130, 148, 150

G

gas stripping, 53, 57, 58, 59, 71, 75, 82
genes, 18, 20, 23, 47, 73, 80
genetic engineering, 18, 20, 62, 85
glucose, 13, 14, 15, 16, 44, 45, 46, 47, 49, 50, 51, 57, 62, 71
glycerol, 24, 45, 49, 67, 133, 134
growth, 13, 43, 45, 47, 48, 71, 107, 108

H

hemicellulose, 33, 35, 37, 38, 39, 40, 41, 42, 78
heterogeneous azeotrope, 53, 62
homogeneous azeotrope, 53, 76
hormones, 4, 8, 19
human, 32, 102, 103
hybrid, 64, 70, 82
hydrogen, ix, 38, 39, 47, 73, 94, 101, 103, 104, 110, 112, 115, 119, 120, 132, 133
hydrogen peroxide, 73, 94, 104, 133
hydrogen-rich compound, ix, 102, 119
hydrolysis, vii, viii, 11, 15, 16, 28, 37, 40, 44, 48, 61, 72, 77, 83, 84, 94, 95
hydroxyl, 7, 12, 30

I

immobilization, 16, 17, 51
improvements, viii, 2, 77
industry/industries, 4, 7, 16, 18, 19, 20, 29, 34, 60, 61, 77
inhibition, 16, 41, 43, 48, 62, 71, 79, 117, 134
inhibitor(s), 16, 35, 38, 40, 42, 43, 44, 45, 68, 76, 84
inoculation, 14, 17, 51
isomers, 3, 21, 22, 110, 111, 112, 113, 114, 126, 127, 128
Israel, v, 27, 86
issues, 20, 30, 56, 62, 66, 70, 91, 96

K

kinetics, 75, 107, 135

L

lactic acid, 47, 93, 94

Lactobacillus, 47, 62, 68, 85
lactose, 14, 45, 48, 50, 79
lead, ix, 102, 115
lignin, 17, 33, 35, 37, 38, 39, 40, 41, 42, 44, 78
lignocellulosic materials, viii, 28, 31, 33, 34, 39, 61, 83
liquid-liquid extraction, 17, 53, 54, 55, 56, 58, 59, 63, 64
liquids, 38, 39, 138

M

magnesium, 9, 10, 16
mass, 11, 37, 51, 87, 88
materials, viii, ix, 28, 30, 31, 32, 33, 34, 35, 37, 39, 59, 61, 65, 83, 102, 103, 125, 128, 133, 140
media, 13, 14, 111
membrane distillation, 53, 55, 68, 80
Metabolic, 17, 24, 46, 67, 77, 83
metals, ix, 102, 105, 112, 117, 120
methanol, 18, 22, 38, 109, 110, 111, 114, 115, 116, 127, 131
Mexico, 27, 32, 33, 86, 89, 93, 96
microorganism(s), 3, 11, 14, 35, 37, 41, 43, 45, 47, 48, 54, 62, 66, 68, 74
molasses, 18, 43, 48, 49, 79, 83, 84
molecular structure, 111, 112, 115, 126, 127
molecules, 37, 41, 105, 110, 112, 115, 116, 135, 137
morphology, ix, 83, 102, 106, 107, 108

N

nanoparticles, ix, 102, 103, 106, 107, 108, 113, 120, 127, 131, 132, 134, 135

O

oil, 28, 73, 88
optimization, 66, 67, 69, 70, 76, 80, 82, 86, 89, 90, 91, 92, 95, 96, 97
organic solvents, 8, 20, 38, 79
oxidation, vii, ix, 21, 38, 102, 103, 104, 105, 108, 110, 111, 112, 113, 114, 115, 116, 117, 118, 119, 120, 121, 125, 126, 127, 128, 131, 132, 133, 134, 135, 137, 143
oxygen, 3, 16, 19, 30, 38, 103, 104, 108, 109, 115, 117, 127, 132, 133, 134, 135, 144

P

pathway(s), 12, 18, 23, 46, 47, 62, 66, 68, 73
petroleum, 20, 28, 30, 60
pH, 11, 14, 15, 47, 72, 127, 132, 134, 135, 140, 141, 144, 145, 146
pharmaceutical, 19, 20, 21
phosphate, 2, 39, 68
plants, 11, 15, 40, 60
platinum, ix, 102, 111, 112, 116, 117, 123, 124, 125, 126, 127, 128, 132, 133, 134, 135, 137, 138, 139, 140, 141, 142, 143, 144, 145, 146, 147, 148
polymerization, 36, 37, 38
pretreatment technologies, 30, 35
principles, vii, ix, 20, 39, 102
production, v, vii, viii, ix, 1, 2, 3, 4, 7, 8, 9, 10, 11, 12, 13, 14, 15, 16, 17, 18, 19, 20, 21, 22, 23, 24, 27, 28, 29, 30, 31, 32, 33, 34, 35, 36, 37, 42, 44, 45, 48, 49, 52, 58, 59, 60, 61, 62, 64, 66, 67, 68, 69, 70, 71, 72, 73, 74, 75, 76, 77, 78, 79, 80, 81, 82, 83, 84, 85, 86, 87, 89, 90, 91, 92, 93, 94, 96, 97, 98, 99, 101, 103, 104, 109, 128, 146, 149
propylene, 3, 10, 55

Pt nanoparticles, 107, 108, 123, 127, 134, 143, 146
Pt single-crystal, 107, 123, 133, 140, 146
purification, viii, 17, 28, 45, 52, 54, 58, 59, 60, 63, 64, 66, 67, 69, 70, 76, 81, 82, 87, 91, 92, 93, 94, 96
purification technologies, 28, 58, 93

R

radiation, 36, 37, 138
raw materials, vii, viii, 19, 28, 30, 31, 32, 33, 37, 48, 61, 66
reactant(s), 18, 104, 105
reaction rate, 38, 105, 108, 120
reactions, viii, 2, 38, 103, 104, 107, 108, 112, 114, 115, 116, 117, 131, 133, 135
reactivity, 37, 107, 111, 113, 114, 120, 126, 127, 133, 135
recovery, 16, 40, 52, 53, 54, 56, 63, 64, 71, 72, 73, 74, 76, 78, 80, 81, 83, 85
recycling, 16, 17, 84, 85
renewable energy, ix, 73, 77, 101, 102, 103, 125
requirement(s), 36, 40, 54, 58, 60, 62, 63, 65, 66, 76
residue(s), vii, 1, 11, 14, 33, 41, 49, 61, 69, 71, 79, 82, 94
resistance, 62, 113, 114
resources, ix, 3, 28, 101, 103
rhodium, 3, 9, 117
room temperature, 10, 112, 117

S

safety, 20, 64, 70, 87, 91, 96
selectivity, 52, 54, 55, 56, 57, 59, 107
separation process, 52, 54, 55, 56, 64, 66, 80, 82, 91, 95, 98
shape, ix, 45, 102, 106, 107, 108, 109
showing, 20, 37, 107

simulation, 65, 75, 80, 87, 88, 89, 94
single crystal, v, vii, ix, 101, 102, 106, 110, 112, 120, 122, 123, 125, 127, 131, 132, 133, 134, 135, 137, 138, 139, 141, 142, 143, 144, 146, 147
single crystals, 125, 127, 133
sodium, 38, 43, 44
solubility, 5, 6, 10
solution, 14, 53, 75, 83, 104, 116, 127, 135
solvents, 3, 15, 20, 21, 38, 39, 40, 47, 48, 49, 56, 63, 68, 71, 77, 82, 85
species, ix, 19, 42, 45, 82, 102, 105, 107, 110, 115, 116, 117, 118, 119, 127, 135
stability, 17, 67, 107, 134, 135
starch, 15, 32, 45, 48, 73
state(s), 13, 24, 30, 39, 40, 55, 105
structure, 33, 37, 38, 39, 41, 65, 77, 103, 105, 106, 107, 112, 113, 127, 135
substrate(s), 11, 12, 13, 16, 17, 20, 46, 48, 49, 50, 52, 71, 72, 73, 76, 79, 135
sucrose, 25, 45, 48
sugar/starchy materials, 31, 33
sugarcane, 31, 32, 33, 76, 77, 83, 84
sugars, 7, 15, 30, 31, 32, 34, 38, 46, 47, 61, 84
Sun, 38, 60, 77, 83, 118, 126, 128
surface area, 38, 41, 107, 122
surface structure, 105, 106, 107, 113, 131, 135
synthesis, 4, 6, 18, 47, 62, 70, 81, 83, 89, 92, 97, 107, 128, 131

T

techniques, 17, 23, 51, 53, 54, 60

technology/technologies, vii, 1, 7, 18, 20, 22, 28, 30, 35, 39, 40, 58, 59, 60, 62, 63, 65, 69, 74, 76, 81, 93, 125, 104, 109, 110, 121
temperature, 5, 6, 11, 14, 21, 22, 31, 36, 37, 40, 41, 55
toxicity, 13, 19, 20, 48
transformation, vii, viii, ix, 28, 31, 40, 101
transportation, 28, 29, 65
treatment, 14, 40, 81

U

USA, 130, 147, 149

V

vacuum, 54, 55, 72, 80
vapor, viii, 1, 5, 30, 65
vitamins, 4, 8, 19, 48

W

waste, vii, 1, 6, 11, 33, 50, 69, 73, 75, 81, 84, 86
water, 3, 7, 8, 10, 19, 22, 30, 37, 38, 39, 40, 41, 44, 52, 54, 55, 56, 58, 59, 62, 83, 115, 117, 119
worldwide, 28, 29, 31, 33

Y

yield, 11, 13, 16, 17, 39, 42, 49, 51, 62, 66, 85, 105, 110